from Estella Leopold

Stories from the Leopold Shack

Stories from the Leopold Shack

Sand County Revisited

Estella B. Leopold

Photographs by A. Carl Leopold

UNIVERSITY PRESS

Oxford University Press is a department of the University of Oxford.
It furthers the University's objective of excellence in research, scholarship,
and education by publishing worldwide. Oxford is a registered trade mark of
Oxford University Press in the UK and certain other countries.

Published in the United States of America by Oxford University Press
198 Madison Avenue, New York, NY 10016, United States of America.

© Estella B. Leopold 2016

All rights reserved. No part of this publication may be reproduced,
stored in a retrieval system, or transmitted, in any form or by any means,
without the prior permission in writing of Oxford University Press,
or as expressly permitted by law, by license, or under terms agreed with
the appropriate reproduction rights organization. Inquiries concerning
reproduction outside the scope of the above should be sent to the
Rights Department, Oxford University Press, at the address above.

You must not circulate this work in any other form,
and you must impose this same condition on any acquirer.

Library of Congress Cataloging-in-Publication Data
Names: Leopold, Estella B.
Title: Stories from the Leopold shack : sand county
revisited / Estella B. Leopold.
Other titles: Sand county revisited
Description: Oxford: Oxford University Press, [2016] | Includes
bibliographical references and index.
Identifiers: LCCN 2015043334 | ISBN 978-0-19-046322-9
Subjects: LCSH: Restoration ecology—Wisconsin. | Restoration
ecology—United States. | Nature conservation—Wisconsin. | Nature
conservation—United States. | Sauk County (Wis.)
Classification: LCC QH105.W6 L46 2016 | DDC 508.73—dc23
LC record available at https://lccn.loc.gov/2015043334

1 3 5 7 9 8 6 4 2
Printed by Edwards Brothers, USA

Also by the author (with Herbert W. Meyer): *Saved in Time: The Fight to Establish
Florissant Fossil Beds National Monument, Colorado* (2012).

Contents

Preface　xi
Acknowledgments　xv

Chapter 1　The Shack Enterprise　1

　Rebuilding
　The Inaugural Visit
　We Meet the Neighbors
　The Shack: Look, Mother, Someone *Lives* There!
　Planting Pines
　The Music of Our Days and Nights
　Our Second Fireplace, a Remodel (1936)
　The Flying Visitor
　Woopsie!

Chapter 2　Winter　49

　Cutting Wood, Banding Birds
　Our Shack Is Vandalized
　The Slough and the River
　Games in Winter
　Cutting the Good Oak

Chapter 3　Spring　73

　Planting Again
　Poco and Pedro

Sky Dance
Warbler Watching
Meat Rock and Calling to the Owls
Goose Music
What Species Do the Deer Prefer?
Road Kill for Supper

Chapter 4   Summer   101

The Rhythms of Summer
Tree House
Leopold Benches
Our Beach
On the Shores of Lake Chapman
What We Found in the Sand Blow
Later Years: Building Trails

Chapter 5   Fall   130

Bounty from Our Shack Garden and Orchard
Carl's Hawks
Hunting Traditions
Early Deer Hunting Near the Shack Property
Dad and Gus

Chapter 6   The Evolving Archery Endeavors   156

Artisan and Archery
Roving and Archery Practice
Mother's Tournament Successes
Lady Diana
Hunting at the Shack and Beyond

CHAPTER 7   THE SHACK LANDSCAPE AND ITS
            RESTORATION: A NATURAL
            HISTORY   176

  The Lay of the Land
  What We Did on the Land: Restoration Efforts
  We Plant an Oak

CHAPTER 8   THE CONTINUING PROCESS OF
            RESTORATION, 1948–PRESENT   215

  The Aldo Leopold Memorial Reserve
  The Bradley Study Center and a Prairie Experiment
  The Leopold Fellows Program
  The Significance of Prairie Building
  The Aldo Leopold Foundation
  Charlie Bradley's Woods and Prairie
  Restored Vegetation Areas
  Oak Forests and Resilient Prairie Plants
  Of Sandhill Cranes and Ducks
  Nina's Phenology
  Other Restoration Projects
  Driftless Area Landowners

CHAPTER 9   THE SHACK IDEA   245

  The Results: A Mosaic
  Starker's Place at Sage Hen Field Station, California
  Luna's Place on the New Fork, Wyoming
  Nina's and Charlie's Place near the Wisconsin Shack
  Carl's and Lynn's Shack in Costa Rica
  My Shack West in Colorado

CHAPTER 10 EPILOGUE: FAMILY AND
            FAMILIARITY   268

Appendices   279

   Three Pet Stories
   Where Did They Come From?

Notes and Sources   299

Index   307

Preface

The Shack is a small barn in south-central Wisconsin that my family retrofitted for camping in the 1930s and 1940s. The Shack has always been the hub of family activity, at least since my childhood. It was also a reciprocal exercise in restoration—every weekend we worked on restoring the land; every weekend it restored us. Dad and Mother took us camping starting at an early age, mostly on weekends. By the time I came along, in 1927, my two older brothers Starker and Luna (eleven and thirteen years my elder) were very experienced. And to a lesser extent so was my sister, Nina (nine years my elder), and my brother Carl (only seven years my senior).

At first, being younger, I was a helper in the family; later I was treated as a full participant in family activities. Dad planned many of these weekend trips, some for hunting and some for the family to go "roving," as he put it, with bows and arrows.

He hoped to find us a place of our own to practice archery and hunt. In the 1930s Dad, a professor of wildlife ecology at the University of Wisconsin, Madison, was already a prominent spokesman for land conservation in the state, though this was well before he wrote his famous book *A Sand County Almanac*.[1]

My father had developed a craving to have his own land to experiment with a new idea: *ecological restoration*. We needed, he said, to find out what the original vegetation had been like in our area and what we could do to bring it back. That, and his desire to have a special place to hunt, led to his purchase in the mid-1930s of an abandoned farm along the Wisconsin River, in the Sand Counties—"the Shack." He specifically chose the Shack land because of its isolation and because this farm was a land of impoverished soil that had become an agricultural failure. In his view this was sick land that needed restoration; it needed to see again the native species that once must have grown here. It was one instance of his larger vision of the countrywide importance of land health and fostering the community of life.

Aldo Leopold came from a close-knit German family in Iowa, and he took after his father in his magnetic attachment to nature and in being a great observer and a creative craftsman. My father was a person of great talent who loved the out-of-doors, and was especially comfortable in the country. While he was a fine naturalist and a professor, it was in the out-of-doors that Dad opened up with his family and had a joyful time. He always looked for the best qualities in people and never looked down on anyone. He was a wonderful father who cherished us, his children, and his students as well. But the love of his life was my mother, Estella Bergere Leopold, of Santa Fe. She was the center of his universe.

Our mother also exerted a powerful influence in our family. As my sister Nina once said, "Mother provided the peace that gave my father the environment for his great thoughts." Mother was beautiful in her Spanish way, charming, gracious, possessed of a great sense of humor, well organized and fun, and devoted to her husband. Their marriage was clearly something very special. When Dad walked in the door at noon for

Mother and Dad in the field. Mother is wearing her silver concha belt. Dad is holding his Fox 20-gauge shotgun. Perhaps in Adams County hunting grouse.

lunch each day, Mother was always there to greet him, and they would hold each other in their arms, in some cases for quite a long time. After a meal they would sometimes hold hands at the dinner table.

They would talk with us children around the table, asking us about our studies. Dad in particular would ask each of us, "Tell me what you found interesting in your studies." In other words, what's new? We were each on edge to have an answer. And at the Shack, it was inspiring to talk through what we were seeing in the field with Dad and what it meant.

From these beginnings, each of us, Aldo and Estella's children, developed a lifelong interest in the natural world and a belief in importance of caring for it. In our explorations we each wandered into different aspects of the ecosystems around us. At the same time we so enjoyed the broad experiences and activities with our parents observing and working with the landscapes at hand. Starker became a wildlife professor at the University of California, Berkeley, following in the footsteps of his father. Luna went into engineering and hydrology/geology, studying erosion. Nina, a geographer, became well known as a lecturer on her father's land ethic. Carl was a botanist/ecologist who went into plant physiology, and later did important work on bringing back tropical rain forest in Costa Rica. I took a path that led into botany, ecology, and environmental history using fossil pollen, and have taught for many years at the University of Washington.

I write now about the experience of growing up in this remarkable family at the Shack, looking back, to convey something of the impressions it made on us; all the good times our parents, Aldo and Estella, had with us, my brothers and sister, and all the times of pain and sorrow, too. I especially want to record the kinds of work we did together that seemed to transform one part of the land and then another. Our experiences at the Shack taught us how to live simply, happily, and responsibly as we were growing up, learning from, and coming to love this special piece of land.

Acknowledgments

On long flights traveling to Wisconsin I was writing a series of stories for my family's grandchildren so that they would have the background to know about our family shack. I thank my colleague Lynn Bahrych for suggesting that these stories might be the basis of a book, and also for her help in starting this endeavor. I am greatly indebted to Jonathan Cobb, formerly of Island Press, who kindly went over early drafts to help me smooth out ideas and text. He has been a very special guide for me, and I admire him for his creative thinking and gentle suggestions. From time to time I was so fortunate to have advice from Curt Meine and Susan Flader, to whom we are all grateful for their leadership as Leopold Scholars, as they brought Aldo Leopold to the light of day well after we had lost him as a father. I was extremely fortunate to have the sturdy support and encouragement of editor Timothy Bent of Oxford University Press, as he, Amy Whitmer, and Alyssa O'Connell were such wonderful editors and overseers of this project.

Biology staff member Kay Suiter kindly turned my very early hand written stories into readable script. I was encouraged when she told me she liked some of the stories. I am grateful to Tom Daniel, Ray Huey, and Toby Bradshaw, leaders at the Biology Department here at the University of Washington for giving me space and an office in Johnson Hall where I could work well after my retirement; that permission and nice facility made all the difference. Right here in my Johnson lab, I had help from Jordan Holley as a lab assistant and graphic artist. This book would never have been written with out the continuing help of my lab manager, Stephanie Zaborac-Reed, whose remarkable talents of all kinds, including computer knowledge, were essential for putting this book together. I cannot thank Stephanie enough for her unflagging support and help. I am grateful to ecologist Peter Dunwiddie for his enthusiasm and for checking out the correct botanical names of the plants we love. I would also like to thank Lisa Doley and Jordan Holly for their help.

At the Leopold Foundation in Baraboo, Wisconsin, I am grateful for kind assistance and support from W. Buddy Huffaker, our president, and for the help with ideas and photographs from the archives by staffers Jennifer Kobylecky, Jeannine Richards, Maria Kopecky, Steve Swenson, Jennifer Anstett, Teresa Mayer, and Teresa Searock, among others. Wonderful people all! It has been a fun project indeed. I thank my local family, Susan and Scott Freeman and son Peter, for their advice and encouragement in difficult times, and for their incredible hospitality; their home is a virtual haven for wandering scientists and friends. The talented Peter Freeman kindly converted my music to script for my tribute to cottonwood trees at the Leopold Shack. I thank my nieces and nephews and their children for their enthusiasm and help as we move forward to take care of our iconic Shack. I would especially like to thank Lynn Bradley Leopold for kindly editing the section on the Tropical Forest Initiative.

Photographs are courtesy of the Aldo Leopold Foundation Archives, and most were taken by A. Carl Leopold, unless otherwise noted.

Stories from the Leopold Shack

One

The Shack Enterprise

Rebuilding

In each person's life a particular place may stand out—a place where one spent a lot of time, a place one grew to love and recall for so many happy memories. Such a place for me was the Shack, on the floodplain of the Wisconsin River. In summertime, standing by the river, it was incredibly quiet, except for the occasional call of a kingfisher. It often seemed that high overhead one could hear a kind of humming. Look up and there were barn swallows turning in the air catching insects. Look down and the surface of the river was always quietly in motion, and rippling against a snag in the shallows.

We were a hunting and fishing family. Although camping on weekends early on became a family tradition in Wisconsin, Dad got it into his head to buy a piece of land of our own on which we could camp, hunt, fish, swim, and study nature and even do some bow hunting. He also had a real itch to practice a new idea, ecological restoration, on his own land. At the dedication ceremony of the University of Wisconsin Arboretum on June 17, 1934, Dad told the audience: "The time has come for science to busy itself with the earth itself. The first step is to reconstruct a sample of what we had to start with. That, in a nut shell, is the Arboretum." He was looking for a place of our own to do just that as well—"a place to show what the land was, what it is, and what it ought to be."[1]

It was in January of 1934; Dad asked an archery friend of his in Prairie du Sac, Ed Ochsner, to help him locate and lease some land near the Wisconsin River. They visited an eighty-acre piece in the south-central part of the state northeast of Baraboo. Dad apparently thought it would fit his purposes. By paying the taxes we could buy the land for just eight dollars an acre. This was still the time of the Great Depression, and because we were living on Dad's very modest teaching salary, he was looking for a bargain—and at $640 he had found one.

The Inaugural Visit

I was eight years old when Dad took us all up to see the land in early spring of 1935. We drove in our old Chevy with Flick, our springer spaniel. The weather was cold and brisk. Leaving the paved road, we drove a mile and a half on a muddy, single-track road that ran parallel with the river. In *A Sand County Almanac*, the book Dad would eventually write about the experience, he called it the "emigrant road."[2] We would learn about the pioneers when we heard stories from our neighbors. We arrived at the entrance to this old farm at the end of a long row of big elms perpendicular to the river. The farmer's access road along the elm row was impossible to negotiate, being muddy and wet, so we drove in a curved arc across the drier soil of the abandoned cornfield, filled with weeds, sand burrs, and old cornstalks down to the only remaining building on the farm, a small barn. That soil now looked like pure sand, though the census data suggest it had been very profitable at one time.[3] Apparently the farmer had grown corn crop after corn crop until he went broke and abandoned the farm.

We got out of the car and looked over the structure. Admittedly, it was a little unnerving, as there were holes in the roof "large enough to throw a cat through," as Dad put it. There was no door. On the dirt floor at the far end of the barn was a pile of manure at least two feet deep and frozen solid. It seems the farmer had kept both a horse and some chickens in that little barn. The interior was bleak, dark, and dirty, but we could see fine heavy hand-hewn beams. I can remember us stepping back from the little barn and Mother asking, rather anxiously, "Aldo, what do you think about this place?" We children were silent, waiting to hear what Dad would say. I cannot remember his words, but he was quite jubilant that he had indeed found an isolated place so full of "opportunities." It had great potential, he said. Then we all began to chatter about this new property.

The family got out the picnic basket Mother had packed, and we ate some cold sandwiches. We walked together upstream on the low hill overlooking the barn and found the foundation of the farmer's house, which had burned down. In the basement of the old house were barrel hoops, some pots, and the remains of an old still. The hilltop overlooked a long slough of the river toward the north and an apple orchard

Rebuilding the shack. The original Shack (barn) before we attached the bunkhouse. Luna is framing the new doorway. Young Estella is on the log watching. Nina is seen on the left carrying a basket. Notice the tin stovepipe chimney of our first fireplace and the fact that the barn had only one window on the west wall.

on the east slope with a few trees. A row of grand red cedar trees stood along the north side of the orchard.

The land was pretty open; you could see for a mile in two directions. The cornfield was full of weeds and sand burrs, and the soil was about pure sand. The little barn sat on a wide sweeping plain, which was actually a high, flat terrace that stood about twelve feet above the active floodplain, dotted with sandbars and willow bushes. The river ran along the margin of the farm to the north. A long ridge perpendicular to the river formed a hill directly west of the little barn. The access "emigrant road" (parallel to the river) crossed the hill, and one could see that red clay marked the top of the ridge to the south (we later called it "Clay Hill") and orange sand topped the ridge to the north, now called "Sand Hill."

We children began to get pretty excited. It was easy to imagine future swimming, tree climbing, hunting, and camping on this place along the river, so far from any development. We even spotted a nice horizontal branch high in the elm in front of the barn, a perfect position for a swing.

We began learning about this region, where, according to author Steve Laubach, the Winnebago Indians had pushed out the Sauk and Fox tribes, and after conflicts with settlers in 1827 the US government had built Fort Winnebago near Portage, Wisconsin. The early explorers described the land as being "a mix of oak savanna, marshland, and forest, with excellent hunting grounds for deer, elk, moose, bear, beavers and ducks."[4]

Among the settlers the initial landowners of the property of our interest were William and Caroline Baxter, who had acquired a large piece of the property in the area after the 1840s and by 1860 were reportedly growing corn and wheat, yielding about five hundred bushels.[5] For a time they ran a small dairy. By the 1870s the fifteen apple trees Baxter had planted were yielding up to forty bushels.[6] Some of these were on the slope west of the Shack site and were still yielding well in the 1930s. Some Baxters still live in the local area. In about 1915, they sold eighty acres to Jacob and Emma Alexander, who grew corn and row crops and raised sheep, cows, and chickens.

The sand blow, looking northwest (1935). The sand blow is a patch of pure sand at the top of the Sand Hill; in the foreground the muddy driveway leads to the Shack. The row of old elms is visible on the right. Cedar trees (*back, right*) were planted by a previous farmer. Carl took this photo by climbing a hickory tree at the gate to the Shack property.

Map 1, showing place names of the Shack area, 1937. Based on sketch map by Aldo Leopold. Dad was especially aware of the two deer trails shown here.

After their house burned down and they were unable to pay their taxes, the Alexanders left the area in 1933.

My father leased the land in 1934 and closed on the purchase of the eighty acres in the spring of 1935. At this time he was calling the property the *Jagdschloss* (hunting lodge). It was not long before he began to talk about making an effort to bring back the native plant species that should be on our farm; it was just a matter of deciding how to do it. First we were going to need to fix up the little barn for camping. He hoped that we all would help.

One thing led to another, as you will see in this account. It was easy to become interested in Dad's ideas about the place.

We Meet the Neighbors

On our early weekend trips from Madison to our newfound land, we would stop in Prairie du Sac at Ed Ochsner's red brick house on the main street that paralleled the river. Ed would come out to talk with Mother and Dad about archery. Often he would give us a big tin pail of honey, usually with the honeycomb in it. We loved that on our pancakes. The smell of Ed's honey warming on a sourdough pancake covered with butter and the chewy taste of the honeycomb on the pancake was unforgettable. And with bacon!

We all loved talking with Ed. In his rasping voice he used to tell us tales about the old log home on the river terrace in Portage. The site of Portage was the place between the Fox and Wisconsin Rivers where the tribes and later the explorers crossed the river divide between the Great Lakes drainage and the Mississippi River. He called it the "Settlement House" and said it was occupied by an early government agent who traded goods with the Winnebago, the remaining Sauk Indians, and the settlers. From Ed's accounts, the area had been generally peaceful. We always cringed, though, when we heard and read about how in 1832 Chief Blackhawk put his Sauk people, women and children, on a raft just below the present Sauk City on the Wisconsin River only to have federal troopers shoot them one by one with their rifles from the south bank. I always hated to think about that tragedy, which is now commemorated by a bronze historical plaque marking the place by the river where the ambush occurred. Chief Blackhawk himself escaped and hid for a while in the hills. We used to pass that site every weekend on our way to and from the Shack. It always gave me sobering thoughts.

Next door and downstream from the land that Dad bought was another farm, also of about eighty acres, occupied by a Mr. and Miss Gilbert. They were roughly in their seventies when we bought the Shack property. They lived in a frame house next to a well-built log cabin, which may have been their first home and was being used as an outbuilding for storage. We liked going down to visit these people, who seemed so elderly to us, and hear them talk about "the early days." They were growing corn and had a vegetable garden.

The Gilberts' farm was situated on a flat floodplain abutting a marsh at the downstream end ("Plummer's Slough"), beyond which was a prominent ridge perpendicular to the river. The place where the river washed against the nose of that cliff was known as Barrows Bluff.

In a crackly high voice, Old Miss Gilbert would lean forward in her rocking chair, point to the muddy road that paralleled the Wisconsin River, and talk about how in the settlement times—the late 1800s—she had watched covered wagons loaded with families and their goods travel this road going upriver. "That was in the old days!" she would say. "Lived here all our lives! But times have certainly changed."

Old Mr. Gilbert, her brother, was a quiet man. He agreed to bring his horse team up to our shack with a plow and plow a food patch for us. The first food patch was developed on the east side of the Shack, where we planted sorghum and hemp. Dad said these grains were great for feeding the birds in fall and winter. The next year we planted potatoes there. A very good crop, satisfying to a hungry family.

The following year another neighbor farmer, Mr. Webster, plowed an area west of our shack in the old orchard we had been fortunate to inherit. We planted a real garden there. The old apple trees continued to bear very tasty fruit, some of which even reminded Mother of the Wolf River apples she ate as a child in New Mexico. In September of 1935, those trees were loaded with huge though sometimes oddly shaped red and yellow fruit.

With a gray beard and a dark slouching hat Mr. Gilbert looked a bit like John Muir. I was in awe of him. He called our place "the Elums," because of the long row of large American elm trees that grew alongside the driveway leading in to the Shack. So Carl also sometimes called this place the Elums. Here was another name for the place.

A barbed wire fence marked the line between Mr. Gilbert's farm and ours. But at least one time, in July 1936, his milk cows went through the fence, came up to our orchard garden, and ate most of our corn crop. We children were particularly

Plowing. Mr. Webster is plowing the garden here, and note the old apple tree in the foreground. We think it is one of those planted by the Baxter family back in the late 1800s. Its apples were very good indeed, and perhaps this tree was a Wolf River apple tree. Looking southeast.

dismayed by the loss of our garden crop, and the family made plans to start building a wood pole fence around our orchard garden.

At the southern edge of our land across the muddy road was Lake Chapman, spanned only by the vertical pilings and cross beams of a former bridge. The lake is an elongated, deep water body about a quarter mile long but only fifty yards wide; at one time this was probably the main channel of the Wisconsin River—or at least a major floodway during high

Mother fishing. We caught some nice fish in Lake Chapman occasionally, mainly bluegills.

water. Mr. Gilbert and other neighbors told us the lake was "not real good for fishing." Even so, in our happy hours of fishing from the pilings, we caught a few blue gills from time to time. Fishing in the river was more productive; there we caught walleyed pike using worms.

In 1937, Dad and Mother's good friends Tom and Catherine Coleman of Madison bought some land across the dirt road from us, and that family has been an integral part of our history since.

The Shack: Look, Mother, Someone *Lives* There!

In the spring of 1935, we brought up a set of tools and began the reconstruction and repair of the barn. I can hear Carl's voice calling, "Hey Sis, let's get ready to go up to the Elums! Hot dawg." We worked on cold weekends with pickax and shovel, slowly digging out the frozen manure on the barn floor and spreading it out where Dad thought the garden would go. Early that spring Dad and Starker and Luna built our first fireplace so that we would have a place to cook. It had a tin smokestack. Then in April Dad, Luna, and Carl started work on a new bunk wing for the building we were now calling the Shack. They brought up some lumber supplies, two-by-fours and stripping. As they broke out the west wall to start framing and building what would be the bunkhouse wing, Carl described a little phoebe that kept flying into the structure and perched, scolding the boys, as she apparently wanted to build a nest in the structure.[7] They poured cement sills for a bunkhouse foundation. After a day or two the rest of us arrived, and Starker, who was working on a game farm in Coon Valley, Wisconsin, came along. They finished the roof with tarpaper, built benches all around the inside of the Shack, and framed the bunks. Next we started repairing the rock foundation with cement we had mixed ourselves using sand from the river bars. Carl's journal recorded that Luna was teaching him to how make and bake sourdough bread, which was "mighty tasty."

As this was the period of the Great Depression, we used recycled materials (e.g., old boards) to save as much expense as possible. A small shed near the Shack provided boards to build the bunkhouse wing. During the week, Dad sent Luna and Carl to the city dump in Madison to look for old windows and a door, as well as usable lumber, and someone loaned Dad a trailer for hauling heavy items up to the Shack. We also walked up and down the riverbanks scouting for beams and boards that might have been delivered by the river.

Rebuilding the Shack. Photo shows Starker carrying the ladder preparing to frame in the bunkhouse. Dad is balancing a new two-by-four for the frame. Note the tin chimney for our first fireplace.

Later Dad wrote an essay about such gifts: "An unpredictable miscellany of floatable objects pilfered from upstream farms. An old board has to us twice the value of the same piece new from a lumberyard."[8]

In April, at Birmingham and Hixon of Baraboo we purchased rolls of tarpaper and roofing nails. The nails had flat heads as big as a dime to be driven through shiny tin circles that served as washers to hold down the tarpaper. We repaired the roof, covering it with the heavy tarpaper and putting black tar along the seams. The cement hearth a few inches above the floor was poured when Aunt Anna, Mother's sister, came to visit in April of 1935. Aunt Anna and I wrote our initials in the corner in the fresh cement that day. A picture shows Aunt Anna, accustomed to the warmth of Santa Fe, looking very cold in the front yard. I'm sure some of these

Rebuilding the Shack: Dad is framing the new window on the south wall, while Nina and young Estella look over the old cornfield toward the southeast.

were day trips. On occasion, though, we camped in the front yard with a large tent.

We were also making furniture for our shack. Built-in plank benches now lined the back and sidewalls. We made two long stools and a three-legged stool handy for cooking and sitting right in front of our fireplace. The sturdy table from the dump was great for serving our meals, as it now had a nice shiny pine-board top sanded and treated with linseed oil. We ripped off the vertical siding where it was weak and used recycled oak boards to make new siding. On the inside we put tarpaper lining, then covered it with horizontal boards to keep the wind out.

The first bunks spanned the entire wing of the Shack, about fifteen feet, wide enough to sleep five on each bunk. The upper bunk was also continuous. We laid red snow fencing for springs across the two-by-four supports, laid canvas on the snow fencing, and for mattresses hauled in fresh hay from Mr. Baxter's haystack across the road from our shack. (We later bought that moist meadow property from Mr. Baxter.) We laid bedrolls out on the hay, and the result was a nice place to sleep.

At the time brother Luna was in college at the University of Wisconsin. Starker joined us occasionally from Coon Valley. Both boys came up when they could to help us. Starker's construction was a privy, built in March 1936 of fresh boards and two-by-fours from the lumber company. After we children had helped finish the first step, digging a privy hole about four by four feet across and five feet deep, Starker framed the privy using two-by-fours to support the sloping roof. Starker came into the yard for lunch that day and announced with his boyish grin, "The privy is so beautiful, it's grander than the Parthenon!" Everyone laughed and went out to see. The name stuck. Of

After we completed the bunks and the bunk wing, we were hauling in a load of fresh hay courtesy of Mr. Baxter's haystack. It served as a fine mattress.

course it's been moved several times, but it still stands and still bears the name "the Parthenon." I was nine years old and was surprised later to learn that there was another Parthenon in Athens. Mother was especially pleased that we now had a proper privy.

Along with that improvement, we had to have a proper floor for the cabin. Dad said we would install a clay floor. Mother was not convinced that this was a good idea. Despite her

The Parthenon. Starker Leopold building the privy. Photo looks north toward the river. Note the horse skull wired to the fencepost on the right. A small house wren nested in that skull almost every year.

doubts, the floor was built and served us well for several years. The clay came from the clay hill where the river road cut through it. With a galvanized-tin laundry tub and shovels, Carl, Luna, and Dad made trip after trip to the road cut, digging out the red clay. At first Carl and his school friend Fred rolled an old two-wheeled cart back and forth to the road cut, where they could get fresh red clay. To make it easier, Dad filled a load in the back of the old Chevy and backed it up to the Shack's door; we all began to shovel the clay into place, starting at the back of the barn. After tamping the clay down with feet, shovels, and boards, we would sprinkle the clay floor with water from our watering can just before leaving on Sunday nights. The idea was to make a hardened clay surface for our floor. It worked pretty well. We could even sweep the floor each morning to pick up the straw or loose dust. The clay floor was completed June 30, 1936.

Very early in our reconstruction period, we had bought a Sandpoint hand pump. Luna, the student engineer, measured the height of the river terrace on which the Shack stands over the mean height of the river floodplain and shore, and estimated it was about ten to twelve feet higher. In the front yard of the Shack, we drove a pipe with a sledgehammer into the ground and hit water at only about eight feet. Then we sawed the top of the pipe off flat and attached that Sandpoint hand pump with a new leather valve. Lo and behold! We primed the pump and pulled up fresh water. Dad had it tested at the university, and indeed our beautiful, sparkling, cold pump water was pure.

Dad got so excited about the idea of our new pump that he had a barrel-making company make some wooden pails held together by metal bands. I always liked those pails because they smelled so nice; they may have been built of cedar. They were enormously heavy to haul around, but they held water just fine as long as we kept them wet (to prevent wood shrinkage).

Once early in July 1935, we were camped at the Shack, cooking supper and considering going to sleep, but the mosquitoes were impossibly thick. The bugs were humming in our ears, biting through our clothes, making life just miserable for all of us. Dad and Mother said, let's go into Baraboo and stay at a motel. This was a novelty. We had *never ever* stayed at a motel (and never did again). Well, this was a sensible solution

Our sand-point pump and wooden buckets in front of the Shack. Note the Leopold bench in the background. Photo looks to the east. Photo courtesy of Susan Flader.

Here is the motley crew lined up for a picture in front of the Shack. The crew are (*left to right*) Aldo Leopold (Dad), Estella B. Leopold (Mother), young Estella, and Starker Leopold. Flicky the springer spaniel is present.

to the mosquito problem, and so that night we all had some reasonable sleep. The very next visit we measured all the windows in order to have screens made. The screens made life far more pleasant during the bug season of early summer.

At about this time Carl lined us all up at the door of the Shack and took a picture. I always thought that in this photo we looked like a family on relief. One day I remember a car went slowly by traveling up the river road, and as the kids looked us over they shouted, "Look, Mother, someone *lives* there!"

By 1936 or 1937, we were coming up to the cabin, which Carl sometimes called "the Shanty"—almost every weekend. During flood season in March we had to walk in a mile and a

Every spring we had floods, typically in April, and very often in the fall there were major floods that made the river road impassable for vehicles. Here Dad and Flicky are carrying our supplies. Camera faces east.

half from the Lewis farm because the river road along the floodplain was not passable. We now had a fireplace to cook on. Even if it smoked, it could still keep us warm. We also had had a marvelous old aluminum kettle identical to the pair of steam kettles that Mother's parents had kept on the kitchen stove in 1900 at the Big House in Santa Fe where she grew up.

That handsome teakettle was one very important piece of cooking equipment at the Shack. Another absolutely central piece was Dad's three-legged Dutch oven. It was ten inches in

My charcoal sketch of our steam kettle, Dutch oven, shovel (to move coals around), and trivet. The trivet is a useful tool when cooking in- or outdoors for placing pots or plates near the fire.

diameter with a cast-iron lid that had a deep rim around it for placing glowing hot oak coals for baking bread. The Dutch held heat from the oak coals so we could cook our sourdough pancakes every morning. Dad and Mother used to cook pork chops, lamb chops, a pheasant, and various other good things like kidney stew in the Dutch oven. The food was delicious. Dad had a metalworker in Baraboo make us a couple of three-legged trivets on which to perch the supper plates to warm near the fire or to put the wire grid that Carl made for toasting our bread a safe distance above the oak coals.[9] The trivets were made of rounds of iron, twisted to make the legs and feet, and a top with an S-shaped crosspiece so the plates wouldn't fall through, plus a short handle. For efficient cooking in the cabin over the oak fire or for cooking outside in the summertime over an open cook fire, the trivets were essential. At night we would light some candles and our glass kerosene lamps. A small shovel for pulling out the hot coals was always hanging by the fireplace.

To cook on our fireplace, we used nesting aluminum pots Dad had purchased during his early camping years from Abercrombie and Fitch. One was eight inches and the other ten inches in diameter, and each pot was about ten inches in height with pail handles and tight-fitting lids. There was a white enamel coffee pot about nine inches tall with a long pouring spout and graceful handle. Thieves took this from us, but Mother found a similar coffee pot of the same shape with blue enamel. We still have that one.

I grew to love these utensils for camp cooking as we shared many a pleasant and tasty meal with them.

Planting Pines

It was Dad's idea that we should pay attention to the plants around us. We should seek out native species to bring to our land, plants that belonged in this part of Wisconsin. How did we determine what species to plant? When we walked up and down along the river looking at the forest stands, we found a fine small stand of mature white pines upstream from our land called Anchor's Woods, after the name of the owners. These pine trees were huge, with trunks about twelve inches in diameter, well-spaced, wide apart and healthy. They were and are beautiful. It is only a small, open grove, but a pretty one. Clearly these pines were a part of the original woods in the region, but none were to be found in our woods, or in other areas nearby. Dad called this grove "an ecological landmark," and he wrote, "It is the nearest spot where a city-worn refugee from the south can hear the wind sing in tall timber. It harbors one of the best remnants of deer, ruffed grouse, and pileated woodpeckers in southern Wisconsin."[10] Their size was evidence that pines had been part of the original forest here. There had been an old sawmill not far from Anchor's Woods, and many large stumps were evident. Apparently in recent years pines had been harvested near our Shack land.

Dad often talked about how Jefferson Davis—later president of the Confederacy during the Civil War, but in 1828 a lieutenant in the US army—had been challenged to go to "Pine Island" downstream from our property to cut pines for the new Fort Winnebago and to float the logs downstream to the fort site near Portage, Wisconsin. This story suggests that white pines were an important species in the woods of our county at the time of settlement. Because there were no pines left in our woods, only a few suggestive stumps, Dad figured that planting some pines was historically the right thing to do. It became part of our restoration scheme to reestablish the piney woods that once grew along the Wisconsin River. Planting native trees and shrubs on our land was an essential element in rebuilding the ecosystem that once flourished here. Probably the original pines in our woods had been harvested.

Early furrows dug on the birch row; Starker and Hammy are examining the tiny pines in our first year of planting. Shack is in the background.

Our lowland woods (which flooded frequently) now comprised mainly soft maple, birch, linden, and elm.

We had two books to guide us: *Gray's Manual of Botany* and *Spring Flora of Wisconsin*,[11] and these told us clearly which species were native in our area and which were not. On the high ground we dug holes and threw in acorns or walnuts so that we could build a diverse forest with hardwoods. In the lowland woods we planted pines on high knolls and along the roadside.

In April 1936 Dad ordered one thousand seedlings of white pines and one thousand seedlings of Norway or red pines and some shrubs from a nursery in Madison, and we began planting. After that it was more like two thousand or three thousand white pines and one thousand red pines every spring. This effort began a tradition: we planted pines every single year after that (see chapter 3). We started by planting the small white pines along the north margin of the old cornfield east of the Shack. After Dad's previous experience in the squared-off forests of Germany, he wanted to develop a simple landscape with conifers around a meadow (cornfield). Dad arranged to have a neighbor plow single irregular furrows to make it easy for us to dig holes in the open soil at the bottom of the furrow.

The first batch consisted of two-year-old seedlings of white pine (*Pinus strobus*) (2/0) and red pine (*Pinus resinosa*) (2/1) wrapped with string in bundles of twenty-five or fifty. When we got them to the Shack, we took some clay from the road cut at the clay hill and made a slurry of red clay and water. We dipped the bundles of pine roots in the clay slurry to coat the roots and prevent them from drying. Then we dug a long, narrow pit under the elm trees at the side of the cabin where the sun could not reach the pine foliage and stashed the bundles in the pit, covering the roots loosely with soil. That way they would stay cool and moist until we were ready to plant them. We were also planting dozens of shrubs, mountain ash, juneberry, nannyberry, cranberry, raspberry, and plum.[12] Every one of them was an exercise in nurturing.

The pines were ordered for the beginning of the university's spring break so that we could reserve a whole week in March or April for planting. We children were always there for the planting, of course, but Dad and Mother often invited Dad's sister Marie from Burlington to help plant, as well as occasional students, and sometimes old friends, such as Ray

Roark; John S. Main and his wife Dorothy Turner Main, the daughter of the well-known historian Fredrick Jackson Turner and mother of my best childhood friend, Lois Main; and Jeanette Tenney and Dr. Kent Tenney, a pediatrician and a neighbor.

Before we began planting each morning, Dad would get all the shovels lined up in the front yard and with a large metal

Sharpening shovels before planting. Dad is sharpening shovels while Luna holds the shovel tight on the bench in front of the Shack. Flicky, our springer spaniel, is in the foreground. This was our first bench, and it was attached to an old elm tree on right. Photo looks westward.

file put a fine edge on them so they would be sharp enough to slice through the toughest sod. We worked in pairs. One carried a pail with a few bundles of pines, fishing out one at a time and placing it in the hole that the shoveler had dug. The shovel person would stamp on the soil so the tree would have good contact with the soil. Each of us helped, and that included Mother. She would hold the tree in the hole while Dad would stomp the fill dirt in place, and then Mother would turn to the little tree and say, "All right, so now you GROW!"

Dad and Mother planting pines, probably on the birch row. Camera faces south, looking toward the great marsh. This area was the original cornfield (ca. 1936). Terbilcox property and hills are in background.

Field conditions during this time of year were typically cool and wet. It was always a good idea to wear field boots, a heavy jacket, and a hat. Gloves did not do much good for the person handling the little trees, but the shovel handler certainly needed them.

The landscape design near the Shack consisted of putting a bunch of pines in a west/east line along the northern margin of the old cornfield at the edge of the high terrace, downstream from the Parthenon. We called that planting the "Birch Row" because of the many river birch saplings there. Dad wanted to keep the view open from the Shack southeastward across our future prairie toward the big marsh (see his essay "The Deer Swath: "When the deer hunter sits down he sits where he can see ahead, and with his back to something.").[13] Though there were other things we might have been doing on the week of spring break in late March or early April, we began to feel as though this planting was creative work. We always had fun.

In successive years, we were planting pines on the clay hill, and on the flat along the row of old "elums," the elm trees that eventually succumbed to Dutch elm disease. We planted dense stands along the eastern margin of our property, close to the Gilbert property and near the river. One of these we called the "Ring Around," because it was here that we had a farmer plow furrows in a spiral. I always loved that area, because we discovered a few baby orchids of the genus *Spiranthes* growing in the Ring Around, and also a dear little purple *Polygala* (milkwort) flower. So pretty! The ground was spongy near these plants with *Polytrichum* (haircap) moss that always looked like tiny miniature pine trees. From that spot one could get a good view of the river.

The first phase of the family's annual pine planting ran from spring 1936 until spring 1939. These were the dust bowl years, and during the first droughty summer of 1936 almost all of the little pines died, even though we watered them regularly. Reed Coleman and I remember that when he came to visit as a youngster we were asked to haul water in those heavy buckets to the pines in the Birch Row. Even so, the mortality rate in the next year was high. The toll was:

Norway pines: 95 percent dead
white pines: 99 percent dead
mountain ash: 100 percent dead
tamaracks: 50 percent dead
grape plantings: all survived but one

We continued to plant thousands of pines the next year, and most of these died, though we did work at hauling buckets of water out to the Birch Row in summer. It was discouraging.

Our family was particularly fortunate that Carl decided to form a camera club in high school and took many photographs during our field trips and Shack trips. He had permission from Mother and Dad to build a darkroom in the basement in our Madison house so he could develop his own film. He built his own enlarger using an old Graflex camera with bellows. When Dad returned from Germany with his brand new Zeiss camera with a 2.8 f-stop lens, Carl's new photos became clear and square with the 120 black-and-white film.

In the third planting year, we ordered somewhat larger pine seedlings and began to plant them in the shade of river birch saplings. This resulted in some improvement in their rate of survival, but many continued to die. By now we knew that we had a real problem with desiccation and mortality, so we began to plant the pines closer together. By around 1940, the summer rainfall had begun to improve—the dust bowl had eased—and most of these seedlings lived and became healthy. Much later, when they all began to grow up to five feet tall or so, we realized we would have to thin the pines. That was in our future.

We established a separate routine of planting prairie species in the summertime (see chapter 4).

The Music of Our Days and Nights

Regular weekend visits to the Shack had become a routine for our family. The Shack property lies about fifty miles northeast of Madison. This meant that every Saturday morning we had a drive that took about an hour and a half, and every Sunday evening we made that same drive back to Madison in the dark. There was plenty to talk about on the drive up. We would always ask for a stop at the Baraboo Bakery in Baraboo, the nearest town to the Shack, to buy some doughnuts, or at Pierce's first grocery store, next to the Baraboo River, where Mother would pick up things she had thought of after departure. On the drive back in the dark on Sunday, we usually took to singing. Often it would be songs that Mother taught us (we were learning the words in Spanish), or cowboy ballads, or songs we learned at school. We tried to harmonize with each other. It was a challenge to try to hold our parts.

On one of these trips, Dad turned to Mother and announced that he wanted a guitar for his birthday. "Oh, lovely," said Mother, "of course!" And Dad continued: "And the first one of you that learns to play it can have it!" *All right*, we said to ourselves. Right away the boys picked out a guitar at Sears, Roebuck for eight dollars, brought it home, and tried to tune it. We sat in the sunroom working with the new guitar. Once they thought they had it right, they began to make up chords. This process went on for a few weeks. Being the youngest, I was trying to absorb all this and to practice a bit when the guitar was not being used. Then Luna announced that we needed to check with the *Encyclopædia Britannica* to see if we had the tuning right. Well, of course we didn't have it right. So the boys retuned the guitar with Mother's piano and began again to make up chords. This time the tuning was more standard. Soon the boys were strumming the guitar and singing along together. From then on, we began to play the guitar and sing in the evenings at the Shack and on camping trips. Dad had worked in the Southwest and for years had heard some fine Mexican music. Mother, of course, loved the New Mexico music she grew up with. She was very musical, and she felt encouraged to sing with us as we learned. One song was "Adelita," which came out of the Spanish-American War. Another was "Cuatro Milpas," about a poor abandoned farm that the singer loved. This part of our Shack experience was a deep and lasting one.

We each did our best to learn to play that guitar and sing. Luna and Starker and Carl seemed to be ahead of us. They had good voices and produced solid chords; Nina played at it only a bit and sang. Carl, a violinist in the high school orchestra, went at it with enthusiasm. As the youngest, I was working to keep up. It was Luna who finally won that first guitar. Not long afterward Starker and Carl bought their own guitars. Carl bought a used Martin with a scarred top, the best instrument we had. Its sound was wonderful. The following Christmas Eve, we sat around the Christmas tree at home in the dark while Dad lit the candles, a German tradition from his Iowa upbringing. I spotted a guitar under the tree. Oh! Since each of the boys already had a guitar now, I thought it must be for me. I held my breath as we waited and waited for the candles on the tree to burn down. Mother and Dad said it was important to admire the tree with real candles. Dad always had a bucket of water nearby just in case. I was fidgeting and waited and waited for what seemed like ages till the candles were blown out and I could get under the tree and see if that new guitar was for me. Well, it *was*. I was simply thrilled.

All this started a family tradition of making music when we were together, including playing and singing together in evenings at the Shack.

Cowboy ballads were a favorite. One of our special guests at the Shack was the geologist Charles Crane Bradley, whose father was a professor at the medical school in Madison. Young Charlie was a great fan of my dad's. I think he had taken Dad's course in wildlife ecology. Charlie played the guitar with skill and knew innumerable songs and ballads, which was a great entertainment for all of us. I learned a few of his songs right away. One summer Charlie had told his parents that he wanted to work on a ranch in the West and be a cowboy for a while. He came home with a number of cowboy ballads. We loved hearing them, and we loved learning them. And we loved Charlie. Years later, he joined our family as Nina's husband.

Our connection to the Southwest was not just through playing its music. In those early years, Mother took us almost every summer or fall by train for a visit to Santa Fe and her sisters at

Above: Carl practicing on the guitar at the door of the Shack, with Estella looking on. *Below*: Starker is strumming on his guitar (ca. 1936).

In Santa Fe at fiesta time mother wore her Spanish costume, jewelry, and tortoise-shell tiara. This photo was when she was about thirty.

the Big House at 135 Grant Avenue, where she had grown up. It gave us a chance to attend the regular Fall Fiesta in Santa Fe, the greatest cultural celebration of the year. Mother and her sisters would dress up in Spanish dresses along with a tortoise-shell tiara and some lovely jewelry. We girls would go to the square (plaza) wearing long dirndl-like skirts with our silver (Navajo) concha belts and lots of jewelry. The men were dressed to kill. Oh, what an exciting social time we had! Here was another chance to hear Mexican mariachis and learn Mexican songs. We thrived on that music. Luna was a special "actor": as he was asked to dress up in his black velvet pants and white shirt and dance at La Fonda Hotel with a beautiful woman in front of an admiring audience. He got a lot of attention.

We all loved these visits to Santa Fe, and some of us had the opportunity to visit the N-Bar ranch where Mother's brothers (the Oteros) raised sheep, near Magdalena, New Mexico.

Our Second Fireplace, a Remodel (1936)

Memorable as those trips were, the Shack was the center of our lives. All was going well with our planting schedules and our efforts to improve our living quarters. However, there was a lingering problem with the first fireplace we had built. It had a sheet of tin across the front, supposedly to release heat, but the fact was the fireplace smoked a good deal. Sometimes, when the wind was just right, it smoked a lot. Luna got it into his head that we should seek some professional advice for a new fireplace. He went to the library and found an old tome by Benjamin Franklin published in 1786 about proper dimensions for a new kind of fireplace.[14] It was full of measurements, especially ratios for the firebox.

Luna thought we might use a single stone slab for the front of the new fireplace we were building. He went to a limestone quarry in Madison and ordered a chunk of rock that was about four and a half feet long and two feet wide. To accommodate these measurements, the slab had to be thick, six to eight inches. Luna and Dad rented a trailer and had the quarry put the slab on the trailer. The thing was immensely heavy, probably heavier than our car. But it was sandy orange in color and beautiful.

The proportions of the firebox, the upper part of the fireplace, needed to be fixed. The boys tore off the top, which was the old tin front of the first fireplace, and recalculated the firebox's correct size and shape. The family started up a cement-mixing operation in the front yard, using sand from the river beach. We built a wooden trough to direct the pump water through a sieve so we could clean the sand. Someone stood at the pump, someone shoveled sand into the system, and someone stirred the cement in a tub at the bottom of the trough. We also lined the firebox and back wall with fire bricks from Goose Lake and inserted right-angle swinging iron bars, shaped in Baraboo, as fireplace cranes, so we could hang kettles over the fire for cooking. These cranes had to be inserted exactly horizontally to swing out and in, one on each side.

Then it was time for the Madison limestone slab. The boys backed the trailer up to the door of the Shack and used logs and two-by-fours to make a runway to roll it in along the Shack's floor to the front of the fireplace. This took much pushing and shoving with pry bars. Luna lifted one end of the

Cement mixing. When we built the revised fireplace at the Shack, we all helped wash the sand from our beach to be used in making cement. Nina and Jean Randolph are at the pump, and I am shoveling sand.

slab with the pry bars and put a two-by-four under it. Then Dad pried the other end of the slab up and added another two-by-four. They added crossbars, also consisting of two-by-fours or beams. Finally, two-by-four by two-by four, they had this huge block of rock sitting level right in front of the fireplace, three or so feet off the ground. They then began to pry up the far edge to tip the slab so it would fit into place on the fireplace. I can remember how dangerous it looked to have this enormously heavy slab perched in front of the fireplace three feet off the ground. Eventually they put cement on the joints and lowered the slab into place. A miracle!

Dad said, "We have to make a nice mantelpiece. How about using cedar?" The boys went out and sacrificed one of the cedars the previous farmer had planted in a row along the edge of the orchard, just above the Shack. Its bark was decorated with little holes made by a sapsucker. Using a crosscut saw and a drawknife, Luna shaped three pieces of the cedar log so that they fit together with tongue and groove. They filled in the top of the firebox behind the cedar log, framing it with cement, and covered it with fresh cedar boards purchased at the Birmingham and Hixon lumberyards in Baraboo. Lo and behold, we had a handsome new fireplace. We finished the cedar wood with linseed oil, which made the wood look so beautiful and red. Dad and the boys had done the heaviest work, but Nina and I, and Mother, too, were right in there, holding pry bars and tools, pitching in and helping with everything we could.

Finally, we replaced the tin chimney with a brick and mortar one. It was a big day when we lit our first fire. We all let out a holler when we found that the chimney drew pretty well. In later years we kept making the chimney top higher by adding bricks to it, but it remains a fine workable fireplace. We sat together every weekend evening in front of that new fireplace talking and singing together, while our Dad and Mother had their *traguitos* (here bourbon and water with a twist of lemon). I loved those times.

When I was in college and taking an art course, I drew a picture in charcoal of our homey fireplace, Dutch oven, and

The second fireplace (ca. 1937). The remodeled fireplace with Luna proudly leaning on the limestone rock across the front. Note all the beams and two-by-fours used to lift and pry the heavy stone front up in place. The homemade kitchen cabinet is seen on the west wall of the Shack.

kettle. I did this while sitting in front of the fire with Mother on a cold, rainy day in 1945. She was knitting while Dad was outside working.

We needed to make further improvements in the Shack. Mother wanted a nice board floor and said that the bunks should be separated with an alley "so we could make the beds properly." The hay bedding needed to be replaced with mattresses. Dad agreed, so we got to work. First, we had to frame

Aldo Leopold and the second fireplace. Dad and the beams used to pry the stone front into position. The hearth of the fireplace is framed in wood and reaches about four feet above the Shack floor. The chimney is not yet completed.

and pour cement ridges (struts) along the clay floor so the boards could lie on something solid and flat. The upper surface of the struts had to be absolutely horizontal.

The cement had to cure and solidify, which took a few days. Then we brought up fresh, wide white-pine boards to make our new floor. Fitting them against the existing foundation was quite a trick, but it got done. We selected a thin Madison limestone slab with ripples, which was laid in a bed of cement, to

Photo of completed rebuild of fireplace. There are two cranes that can hold kettles and/or a Dutch oven over the fire. The kitchen area is on the wall to the right of the fireplace.

go inside the front door. This slab could catch mud or water from the boot traffic of incoming visitors. It worked very well. We also laid smaller slabs of rock in the soil outside in front of the door, so we could knock the mud off our feet before entering. We finished the beautiful new floorboards with a combination of linseed oil and paint thinner. Dad always used linseed oil. It was his favorite finish, and we used it for the cedar logs and boards of the new mantelpiece. The oil made the wood shiny and slightly darker, but allowed it to retain its nice red color.

The Shack was becoming more comfortable now, and easier to keep clean. We used calcimine to whitewash the walls and ceiling. Calcimine paint is basically chalk in a water base with a minimal amount of binders. Not expensive but very white, and it works.

Up until that point we had used chiefly local materials as much as we could aside from the floorboards and the two-by-fours bought at Birmingham and Hixon. During this time we made cement with sand from the beach, and boards and planks recycled from wherever we could find them on the property.

In about 1941, it was time to replace the tarpaper roof with something more substantial. Dad ordered a bunch of cedar shingles and hauled them up on a trailer. I can remember that this was Mother's Day, a sunny day in May when the first birdsfoot violets had come into bloom right in front of the Shack. Our friends Kent and Jeanette Tenney came up to help us nail down the new shingles. They were a lot of fun, and Kent was a fine carpenter.

The Tenneys arrived at the Shack carrying a beautifully made pine ladder of two-by-fours with little pine crossboards set in notches on the top surface. The tops of the ladder were rounded and smooth to the touch. Kent brought this to Mother and Dad saying, "Estella and Aldo, now the ladies can get into the upper bunks easily!" The ladder remains in the Shack to this today. They also brought Mother a Mother's Day card with a wire hanging on the side. When you ran your fingernail over the wire a sing-song female voice called out: "Good *morning*, Mother Dear!" We kept repeating that phrase in that voice, then and in the years ahead: Good *morning*, Mother dear! Mother always responded with a cheerful laugh.

The Flying Visitor

One day in the summer of 1937, while the family was working on the Shack, we heard an airplane flying right overhead. We all stepped outside to look at the novelty, and there it was, a light plane buzzing back and forth over the Shack area. Carl, Nina, and I ran up the hill near the old foundation so we could see the plane better. As it passed over us again we could see a pilot with sunglasses but could not imagine who this might be. The only person we knew who flew an airplane was Charlie Bradley, the young fellow who had been to the Shack to visit Dad (and, it turned out, he and Nina had been dating that season). "Well, Nina," we said, "maybe that *was* Charlie Bradley!" Nina said, "Oh my gosh."

The pilot stuck his arm out of the window, waved, and dropped a small parachute with what looked like a little red film canister tied to it. To our dismay it landed right in the top twigs of a young cottonwood tree. There did not seem any way to reach the parachute without chopping the tree down, until, that is, we brought the ladder up from the Shack. Nina was getting more and more curious about what the mystery pilot might be sending down from the sky. She and Carl told me, "Estella, you are little enough. You climb up the ladder while we hold it and see if you can reach the parachute." I did but could not reach it. So they went down to the Shack and brought up a long bamboo fishing pole. With that at the top of the ladder I was able to knock the thing loose, and it dropped to the ground. My siblings helped me climb down the ladder, and then we went to see what the parachute brought us. Inside the capsule Nina hopefully opened the little red film canister, and we found a paper note that read, "Hi! From Charlie Bradley." After all that, this was the message! Nina was nonetheless impressed, and we all thought the event exciting and funny. Nina was particularly pleased that Charlie would go out of his way to do this. The event was noted in the journal we always kept at the Shack.

Later in the spring, Charlie called Nina and asked if she would like to fly up to the Shack with him and have a little picnic. "Why, yes, I'd love to, but there is no landing field nearby," Nina replied. Charlie said, "No matter, I spotted a good flat alfalfa field near Joe Lewis's farm. We can land there and we can walk in."

So one afternoon Nina and Charlie flew from the Madison airport up to Baraboo and landed in the alfalfa field of Joe Lewis's farm. They carried a basket containing steaks and such to the Shack, and built a fire in the outdoor fireplace. When dark clouds blew in and the winds came up rather fiercely, Charlie became worried. "I think I'd better run back to the plane and tie her down." So he took off at a trot, secured the plane, and came back as fast as he could. Nina got the steaks cooking promptly, and they had a good meal before it was time to head back to the plane. Nina described the return flight as fun and a bit scary, but they got home okay. It was a novel experience, one she never forgot.

Later on Nina got engaged to a wildlife student, Bill Elder, and Charlie married Maynie Riggs, one of the gals in his singing group, the well-known Tudor Singers.

Woopsie!

This story has to do with the new remodeled fireplace. Whenever something fell or got tipped over, Mother would cheerfully say,

"Woopsie!" with special emphasis on the first syllable. And then we would echo "Woopsie!" That became an automatic family response. We all used to like to tell this story to our friends and to other family. When we did, Dad would laugh and laugh. Mother would giggle. We all enjoyed the moment.

Once we had our new fireplace in the Shack, there were still times when the wind direction was just so (from the east) and the fireplace would begin to smoke. The only way to lessen the impact was to open the front door partway, though in doing so

After completing our remodeled fireplace, Dad was finishing our brick chimney. Note that the Shack roof is composed of thick tar paper (shingles came later).

we lost a lot of heat. Dad had the idea that perhaps we should slightly raise the height of the brick chimney.

One fine summer weekend we decided to "fix" the chimney a bit. This involved getting the new bricks into a tub of water to soak so they would meld with the wet cement. We got all the materials together, including a specially mixed batch of cement and sand, and put the ladder up against the outside back wall of the Shack, against the chimney. Dad was high up on the ladder so he could reach to top of the chimney, holding a trowel and a bucket of wet cement. I was in the middle part of the ladder, prepared to hand bricks up to Dad. Carl was on the ground with a tub full of wet bricks.

Carl would hand a brick to me, I would hand it up to Dad, and Dad would take the brick, put wet cement in place on the chimney, and set the brick in place. This was going pretty well. Dad called for a new brick. Carl grabbed one from the tub and handed it to me. I took it and handed it up to Dad. This time Dad dropped the brick. It fell to the ground with a whomp! My voice was in there first with a "Woopsie!" that was about a third above middle C. Carl came in next with an alto "Woopsie!." Then Dad's voice chimed in at an octave below, proclaiming loudly "Woopsie... *Goddddd* daaaammit!"

One of our favorite stories.

With our new fireplace and all the other improvements, we had a fine place to camp. We had become very fond of the Shack, and our visits continued to be as regular as clockwork: season in, season out, each with its own rhythms.

Two

Winter

Winter at the Shack was always a great time, and some weekends it was a big challenge just to get in. After a good snowfall we would park near Mr. Lewis's farmhouse and ski in the mile and a half, carrying our grub. We have a picture I especially love of Mother skiing through the woods, wearing her denim skirt and winter coat. What a great sport she was! And she would holler "Whoopeee!" while sliding down a short terrace in the woods. We were proud of her. Skis were not much in those days—just two waxed boards with a leather strap. But they were better than walking, and fun too.

Passing through the snowy winter landscape was always, in Dad's words, a "search for scats, tracks, feathers, dens, roostings, rubbings, dustings, diggings, feedings, fightings, or preyings collectively known to woodsmen as 'reading sign.'"[1] We could often see many of these signs on the snow. I can remember skiing through the woods with Nina one morning

Three of us skiing along the emigrant road. The snow was so deep we had to ski in and out, carrying our grub in backpacks.

after a heavy snowfall and seeing little "bursts," places where a partridge or two had spent the night in a snowbank and then burst out in the morning to feed. If one wonders how our songbirds survive a cold snowy winter, the answers are revealed on a fresh snow surface: the prairie plants hold their seed pods up away from the snow, and the songbirds land on these dark stalks and remove the seeds. Their dear little tracks show where they were picking up seeds. A way to make a living in winter.

Mother skiing on the old emigrant road. See the top of the Sand Hill on the left; the camera points eastward toward the "Good Oak," on the top left of the road cut. Mother mostly wore skirts, not pants, at the Shack.

Cutting Wood, Banding Birds

For our wood-gathering efforts, our tools were the two-man saw, a double-bit ax with an extra-long handle, two regular axes, a heavy sledgehammer, and two iron wedges. Some of the logs we cut in the woods, though of fireplace length, were too big to carry, so we would split them right there before loading them on the sled.

Our favorite place for the cutting operation was west of the Shack, down the slough and bearing south at what we called the "branch slough" and "the fallen bee tree." Our dog (then Flicky) was always running along with us. We did not know

After a light snowfall on the slough it was entertaining to read "sign" in the snow. Here Dad was observing tracks of deer and rabbits and some bird sign.

the big dead oak we were cutting into was a bee tree, but we sure found out. We cut off a slab of the downed trunk, found the honey, and could not resist taking part of it. Dad said we should put the slab back in place, which we did, and fastened it so the bees there could make it through the winter.

At home the boys had designed and built a large, heavy-duty sled for hauling wood. The runners were two-by-sixes

Dad pulling a load of wood near the "bee tree" to get it on the ice of the slough for hauling eastward.

with steel straps screwed on the bottom. Crosspieces were sturdy one-inch-thick boards, but with two-by-fours at the front and back for strength. All were screwed into the top of the wood runners. Luna braided a thick rope around two huge hardware rings mounted on the front crosspiece for hauling the sled with a lead rope.

The wood-cutting crew was usually Nina, Carl, Dad and Mother, and I, with occasionally Luna and sometimes Starker, who were now away in graduate school. On a bright winter's morning we would all leave the Shack with our lunch and tools on the sled, walk down to the slough, fresh snow crystals

Carl Leopold and Al Hochbaum cutting oak at the bee tree on the upper slough. This was the famous two-man saw that we used regularly.

gleaming on the ice and lightly covering the branches. As we went we could usually see in the sprinkle tracks of deer or muskrat and hear the blue jays calling in the bright sun.

Turning south into the branch slough toward the bee tree we started to plan our cuts. As we turned the long pieces of wood on end and were splitting them with wedges, chickadees were all around us, looking for bugs, ants, or bees to emerge from the cut wood. It took us many weekends to finish cutting the fallen logs near the bee tree, so the chickadees would be waiting for us when they heard the ax working.

Carl and Flick getting ready to haul a load of wood down the slough to the Shack.

Someone would build a fire nearby for our picnic, and it offered a way to warm our hands. Work progressed, and we took turns on the two-man saw. Sounds of the maul beating the wedges into the good oak were followed by the crraaakk and ting as the oak split open and the wedge flew to the ground. We would carry or slide the splits over to the sled and gradually fill our load. The vinegary smell of acetic acid in the air was of freshly split oak.

As the work on the two-man saw progressed, Mother would call out, "REST!" She wanted Dad to take a break. He called her his "chief sawyer" (see "The Good Oak": "Rest! Cries the chief sawyer, and we pause for breath").[2]

Mother or sometimes Carl would be brewing hot soup or tea on the fire, and we would all stop to make sandwiches. By lunchtime we would be pretty warmed up, and we'd all find a place to perch on a log in the sun, but near enough to the fire to enjoy our vittles. The brave little chickadees flitted round the split wood, still looking for bugs, ants, or whatever to eat,

Carl took this photo of a chickadee at our woodcutting station in the Shack woods. The birds were scavenging for ants and bees on the freshly split oak.

and talked to us. When we offered them some pieces of bread they became excited and starting calling. Carl then would pick up their song, making a whistle in his teeth so he sounded exactly like the chicks, calling "spring soon" back to them. He would hold out some bread crumbs, and often one would alight on his hand. When a sled full was had, we would lash the splits down with a rope, pile our equipment on, and begin hauling our sled back down the slough, usually two people

Hauling our fireplace wood on the ice back down the slough toward the Shack. The front sled is my toy sled, being pulled by Mother and me, while Nina and Dad are pulling the new homemade sled.

pulling the load. That was not difficult work when the ice was smooth. But when we got to the downstream end and were pulling across the ground, it took all hands to drag and push the load upslope to the top of the terrace, around ten vertical feet, then overland to the woodpile area at the Shack.

Dad was always particular about how we should pile the wood. I remember Nina telling me that she had piled a whole load piece by piece onto the woodpile, and when Dad got there he took each piece off and carefully repiled it, fitting each

Woodpile at the Shack. Carl is emptying the wooden sled to build the woodpile stacked between a couple of elm trees next to the Shack. Camera faces northwestward toward the top of the Sand Hill. Several mature cedar trees are evident on the hill, planted along the orchard.

piece compactly onto the pile so it took up a minimum of space. We learned how to pile neatly so it didn't look messy. Some of the pieces had to be split again in the yard, which would bring in more chickadees and woodpeckers. There were ways to split the oak so you didn't get your two wedges stuck, and by trial and error we discovered them.

Part of the delight in our winter work was getting to know the chickadees. We talked to them. On the big, heavy oak-slab table which we kept outside and then used to drag into the Shack for safekeeping on Sunday night we put out bird feed, and usually on the elm tree a suet holder. We always had some takers—chickadees, downy woodpeckers, nuthatches, and occasional hairy woodpeckers. We wanted to get to know our local winter birds, to get an idea of the population that frequented our Shack area. This meant banding them. We loved these birds; they were such great company. Some we got to know very well indeed.

A simple trap for catching birds in order to band them is a shallow rectangular wire-mesh-frame open box, hinged at one end. We would prop open the trap with a stick that had a string attached to it. The string went through the cabin window, from which we could observe the birds and pull the string at the right moment. Another contraption was a standard rectangular bird trap made of hardware cloth with a cone-shaped opening and two square compartments. We'd place suet and seeds at the cone-shaped entrance. The birds could pass into the first and then the second or end compartment, where they were trapped and where we could open a little tin access door to put our hand in to catch the birds for banding. We kept careful records of the US Fish and Wildlife band numbers in the journal, and occasionally used a color band as well. Dad wrote a great tale about one of our oldest chicks, who lived about ten years, and we knew him well—65290. He wore a green color band and Dad named him Greenhorn. After a while we could just open the trap door, say hello, and let Greenhorn out of the trap, as we knew him on sight. Clearly Greenhorn did not mind being trapped, as he could always get a free meal. Suet tasted good on a really cold wintry day.

To hold such a tiny, soft breathing creature in one's hand, delicately read the band on his leg, and then let him go was a humbling experience. "To band a bird is to hold a ticket in a great lottery," Dad wrote in his essay "65290."[3] These tiny birds were friends, and we had plenty of company. We loved their calls, "chipper chipper-dee dee dee dee" and "spring soon!"

The real payoff came when we would build up the fire of excellent dry oak and bask in the heat from the fireplace. This was when we could pay homage to the great oak trees that had harnessed the summer sun and now were warming our Dutch oven by the fire. It might soon be time to tune up the guitar and think about supper.

Our Shack Is Vandalized

One cold February weekend in 1939, after our second fireplace was completed, we arrived at the Shack to discover that someone had broken the padlock and gotten inside. It was terrible to walk in and find the utensils and Mother's jam spilled all over the floor. They had taken the large hand drill and cut round holes in our handmade furniture and benches. They probably had found our parents' bottle of bourbon, which likely made it worse. They broke our dishes, glasses, and lamps. They even took an ax to the lovely hand-carved part of the cedar fireplace mantel and cut holes in our coffee pot and cooking pots. It was horrible. We all quietly walked around in the Shack, keeping to ourselves, looking at everything and crying—yes, crying. It all was such a dreadful experience! Pretty soon Dad said, "Well, it is wonderful to see how important this shack place is to all of you! I am impressed by that." There followed all kinds of hugs and a few quiet swear words under our breath.

Then we began to notice they had stolen a few things, among them a pair of brown and white saddle shoes, popular at the time, that belonged to Nina. And some tools had been stolen. I think on our way back through town we reported the event to the sheriff. In any case, a couple of weeks later the sheriff called Dad and said they had found Nina's saddle shoes at Mr. Green's place, where there lived a couple of boys who had apparently robbed us. When they confessed to this, the sheriff put the two young men in jail.

It was the darnedest thing. Each weekend when we would drive in we would pass by the front of the red-stone jailhouse that faced Baraboo's main street. You could look up and see the windows of the jail house with bars, even see the men sitting behind the bars. After about four weeks, Dad stopped and went in to the sheriff's office and asked him to release the two young men who had broken into our shack. Dad said that they had probably paid enough in loneliness for their deeds. The sheriff consented and the men were freed. That said a lot about Dad. Just one instance of Dad's compassion and kindness.

The Slough and the River

Parallel to the Wisconsin River and running eastward is the long slough, or floodway, that carries water from the river during spring floods but in summer is not connected to it. The slough is a delightful feature of the land here because it forms a long, narrow, shallow lake that runs along the base of the sand hill right behind and north of the Shack. One can walk out in back of the Shack, peer over the edge of the terrace on which the Shack sits, and there is the slough. In summer we can put the duck boat in the water there and paddle upstream for a hundred yards or more, in this nice, quiet body of water. One could perhaps flush out some ducks and admire the wall of willow bushes on the right and the bottomland woods on the left.

In the winter this wonderful sluiceway freezes solid and often is good for skating. One time in the 1940s, I carried my skates out back and started upstream along the long runway of ice. The snow crystals on the trees leaning over the slough were sparkling, the ice smooth and transparent, and I could see clearly the mud bottom about three feet down. It was a crisp morning. The chickadees were calling, the sun was out, and it felt good to be skating over that glassy surface.

Then a wonderful thing happened. I looked down and beneath my feet was a muskrat, swimming below the ice. I could see him shoving himself forward with his legs, while wiggling his tail. The muskrat was moving swiftly. I was impressed that he was so competent and fast. I adjusted my speed to match his, and we moved along together over the clear ice mirror. I let out a cheer and felt a sense of delight that I could join this vigorous fellow in enjoying the slough. I do not know how muskrats manage to swim such a long distance without stopping for air, but this one did, and beautifully, or so it seemed. I calculate that we moved together about ten yards before he suddenly dashed sideways and disappeared toward the left bank. I was sure this fellow had other goals in mind than leading me along the ice pathway. Along that slough, there were occasional muskrat houses built of cattails, aquatic plants, and mud. One of these homes was probably his. The whole experience gave me a feeling of being *with* someone on my skating tour, in the company of an able swimmer who knew where he was going, Further, it was someone who lived here, and who knew the area well, we can be sure.

When I returned to the Shack, skates in hand, I recounted the experience to Dad and Mother, who were sitting in front of the fireplace. They were both entertained by my story. Dad and Mother smiled and remarked that I was lucky to have that opportunity to go skating with a vigorous muskrat. The slough was not only the muskrat's highway; it was ours, too.

In the spring and sometime fall floods, the river pours water though this sluiceway. At that time, there is typically a real current of water coursing eastward, connecting the slough with the Wisconsin River. On the river side of the slough is a large sandy, flat "island" with some tall cottonwood trees, river birches, and a lot of open grassland. On the slough side of the "island" were thick border of willows. As a youngster I used to go over to that "island" and play in the sand or sit by the river. We called it an island even though it was wetly connected to the mainland by willows except in floodwater times. Then it really was an island. On the river side it really was a good place to fish. Because I was frequently there, and because no one else in the family seemed ever to go over there, I called this place "my island." I often played on this island; I bonded with this landscape and the cottonwood trees. I loved being there.

The island was not accessible on foot when the floodwaters connected the slough with the river, as water flowed through a thick stand of mature willow shrubs. I was learning how to use the ax and spent an inordinate amount of time cutting soft

Estella Jr. fishing on the "island" in summer. Photo looks downstream on the Wisconsin River.

maple branches and poles to construct a "bridge" that you could walk across through the willow patch (and not get your feet wet). I would cut a tripod of soft maples, usually two or three, with crotches, so they would provide support. Then across the top of two of these tripods I would lay a thick, soft maple tree log. With a staff or pole of some sort I could walk across that bridge to my island during high-water time. My bridge was usually about twenty feet long. After all that construction and cutting of willow brush to open a clearing for a bridge, the flood often swept my structure away during the next season's high water. I was stubborn, though. I think I was using this occupation (bridge-building) to improve ax skills, so the activity went on for a few years. My family laughed at me, but I did not mind.

One time in late winter, when I wandered down to look at my bridge, I heard the most monstrous noise. It sounded like someone was breaking a board, over and over. Craa-ck, craaack! I was mystified and anxious to find out what this was. As I walked across my island toward the river, the noise got louder and louder.

As I approached my island's far shore, I could see that something was moving. My first thought was that a deer was caught in the ice. But as I got closer, I could see the river ice moving downstream. Masses of it. The loudest noise came from upstream to my left, and I headed in that direction. Just as I got opposite the lower edge of another island (which we call Gus's Island), I was amazed to see the river ice driving itself against the north embankment of my island. The ice was bending back and pushing straight upward, to a height of about twelve feet above the ground I was standing on. It climbed upward, cracked, and fell back on itself. Over and over and over, as if on a great conveyor belt. I stood, transfixed, as the detonations were thunderous and the ice cracking almost deafening. The sheer power of what was making that ice, about six to eight inches in thickness, soar and crash in an on-moving stream, was breathtaking.

At the position of Anchor's Island, the river was very wide. The ice between that island and my island was moving slowly downstream with no place to fit as the river narrowed just there. So here the ice was climbing the bank, shuddering, breaking, and falling back on itself—as if on a conveyor belt. The entire river ice was moving like a white blanket.

What a strange and exciting experience it was, to witness this power on the move and crackling with incredible strength. It gave me a real respect for that river. I remember writing about the thrill of witnessing its power in my ninth-grade English class.

Apparently, a year later Dad had this same experience, witnessing the ice layers climbing the shore.

As always, we were on the prowl for valuables that the river would bring us, such as beams, hand-hewn logs, or two-by-fours that might be material for a new bench or repairing the Shack. One cold winter day we spotted a hand-hewn beam about twelve feet long that had got caught in the willows by the slough during a flood. As Dad wrote, "Our lumber pile, recruited entirely from the river, is thus not only a collection of personalities, but an anthology of human strivings in upriver farms and forests."[4]

On a cold day when the slough was still frozen, a group of us walked down to where the beam was and began to pull it through the willow thicket. Bordering the slough, the thicket was dense and the stems were close together. Dad and Mother, Carl, Nina, and I were all trying to extricate the beam. And of course the dog was along to investigate.

Finally we had moved the beam through the worst of the willows. Dad hoisted one end on his shoulder, and Carl, ahead of him, did the same, and we began walking slowly downstream over the ice toward the Shack. We were about in the middle of the slough when Carl suddenly dropped through the ice and found himself waist-deep in cold water! We all shouted, and Carl quickly dropped the beam to the side. For some reason Dad thought this was really funny, and he started laughing and laughing. Mother was shouting "Carl, can we help you? Oh my Goodness!" or some such words. Carl was swearing, "Damn!" He was maneuvering to sit on the edge of the ice and was trying to stand up. Somehow by leaning on the beam Carl was able to lift himself up and stand on to the ice. Nina and Mother and I cheered. We all pulled the beam along on the edge of the ice toward the Shack. We were in a hurry to get Carl in front of the fire place to warm up, strip, and dry off. By now Carl was laughing too, a good sport about it all.

Games in Winter

Nina, Carl, and I used to play outdoor games to amuse ourselves. One was to go out at night in the snow among our growing pine tree saplings and play hide-and-seek. The little trees were only a bit shorter than we were. It was a simple game, but the dark made it challenging and fun.

A game Carl was really good at was tracking. Nina and I would wait on the sand hill while Carl would walk down into the woods. We'd give him about fifteen minutes, then try and follow his tracks in the snow. He was amazingly hard to find. He would walk forward, stop, and step back in his tracks, so we would come to a dead end. Then we would backtrack to see where he had left the trail. Sometimes he would climb a leaning tree and jump off at the far end so his tracks would be nearly impossible to find. Then it would be Nina's turn and then mine.

One adventure I will never forget was a ski trek Carl and I made down by the river at our swimming hole. When Carl announced he was going to ski across the Wisconsin River, I was very frightened and begged him not to do it. Actually, I pleaded with him not to do it. He smiled in his determined way, and said that "it would be fine," which I did not at all believe. So off he went across the river ice, which was patchy with snow. I held Gus the dog. I began to pray, though by that time I had abandoned the Catholic religion; I prayed, and I prayed some more. My eyes followed Carl every minute, as that was the only way I would know if he was okay. This was one of the worst experiences for me of my time at the Shack. The whole trip took a couple of hours. I was waiting at the river's edge, in the cold, terrified. To this day, I do not know how he managed to avoid hitting an opening in the ice—sheer luck.

I chastised Carl royally when he returned, but he just smiled at me and said it was exciting. Of *course* it was exciting. "Wild and crazy" is a better way to describe it. That is one big huge river, full of power, driven by different currents, and no friend to anyone who falls through the ice.

Estella Jr. holding Gus while observing the ice that covers the Wisconsin River at the site of our swimming beach. This photo looking upstream was taken during the previous year (before Carl skied across the river).

Cutting the Good Oak

Growing on the Clay Hill (a small moraine) where the river road crossed it was a large dead oak with a trunk about thirty inches in diameter. This tree had been hit by lightning about a year before, and was now leafless and dry, ready for cutting. In his essay "The Good Oak," Dad noted that after we felled it we counted eighty growth rings on its stump.[5] He went on to suggest that the tree in its youth might have gotten started at a low in the rabbit population cycle (the connection being that when rabbits are abundant, they tend to chew off the baby oaks), a matter that I liked to think about much later when I was in college. Because the tree was eighty years old in 1946, it probably had witnessed the passing of the immigrant covered wagons that Miss Gilbert had described to us, as the wagons went upstream on the river road to settle in land north of here. One of those wagons could even have carried the Muir family, which had come over from Scotland in the 1800s to settle in the Midwest; John Muir's parents did that before they settled in Marquette County, a few miles from Portage.[6] That was a nice thought.

My parents made plans to cut the old oak on a weekend in late winter in March, but it happened to be a week when about eight inches of new snow had fallen. Dad had invited a wildlife ecologist and visitor from Alaska, Dave Spencer, to join us on that weekend. Dave had brought along his skis, and he and I were planning on going cross-country skiing at the Shack that weekend.

That Saturday was fresh, clear, and sunny. Large beautiful snowflakes clung to every branch and twig. The sky was blue and the air was fresh and cold. Mother, Dad, Dave, and I walked to the old oak tree carrying our long two-man crosscut saw, a smaller bucksaw, an ax, and some other tools. The blue jays were calling at us.

Dad looked over the tree and said to me, "Baby, we will want to drop some of those lower limbs before trying to cut the tree. You climb up there and we will hand you the bucksaw so you can start this." I had just settled myself in the tree when I heard a swishing noise. We all looked up into the blue over my head, and there, just above us, were four swans low and in formation, flying toward the river. I held my breath as these beautiful white glistening birds sailed quietly right over us. Their wings were swishing. I grabbed hold of a branch and looked straight up at them. We all watched in awe as these gorgeous large birds sailed quietly above us. It was a moment of reverence for all of us; a heart-warming moment. How very wonderful! We all chattered about this marvelous omen, and those graceful birds, introducing us to the coming of spring. They seemed like a symbol of something wild and free.

Then we concentrated on downing this huge tree. Taking turns at the two-man saw, we eventually dropped the tree, but not without a great deal of effort, as oak wood is dense and hard to cut. Periodically Mother would shout out "REST!," the signal for Dad to slow down and take a breather, and for someone else to take his place for a time. Mom was very protective of her husband, in a very dear way.

It was around two weeks after that when I came downstairs early one morning at home in Madison and greeted Dad, who had been up for hours. He was drinking coffee. "Want to see something?"

"Yes, sure, Dad."

He handed me a few yellow pages covered with his tiny handwriting.

"Take a look, Baby."

There it was, the essay he called "The Good Oak." I sat with him in the dining room reading the story and enjoying every word.

"Oh, Daddy," says I, "this is so wonderful! Would you allow me to try and type the first version of this essay?"

He agreed and I went right at it. I still have the yellow pages (well, I later put them in the custody of the archives). I have always loved the beginning sentence of that essay:

> There are two spiritual dangers in not owning a farm. One is the danger of supposing that breakfast comes from the grocery, and the other that heat comes from the furnace.[7]

I was so proud that he let me type the first version of that lovely essay. And so this part of the *Sand County Almanac* was born.

Dad never did this kind of writing at the Shack. He did it either at home or at his office when he had time. His writing at the Shack was keeping up the Shack journal, recording what was happening, what we were seeing, what the weather was, and what we did. What value that record has to us now!

Well, of course, we reaped the harvest in calories of heat after the good oak was cut up and split and piled, when we burned its wood in our fireplace. We remembered fondly together the March weekend of that snow and the sight of those stately swans flying gently over us. They were lovely, wild, and free.

I have to explain that my father never called me by my first name (Estella). To him there was apparently only one Estella, and that was "his Estella." In his view I had the name "Baby" from the very beginning. I have to further explain that when I was in grade school the nuns could not pronounce my name, which used to be Eloisa Leopold. They called me "Eloise." I consulted Mother and asked if I could use her name, Estella Bergere Leopold (I liked the Spanish Estella), and she said yes. So I called myself by that name ever since, and I officially changed it in my high school years.

Three

Spring

Planting Again

Spring always seemed to begin for us with spring break, when we had a whole week to be at the Shack and do the planting together. Spring is such a special time, with the buds bursting and the early flowers opening. Ever since we started planting in the spring of 1936, we always looked forward to the project, though it meant a fair amount of work, and we always had such a marvelous time. The preparations each year were considerable. Mother and Dad would sit at the dining room table in Madison with a list and plan what kind of meals we might like to have up there and what supplies would be needed. Dad would order in advance thousands of pines from the Conservation District. He ordered at least two-year-old seedlings, usually at least two thousand white pines and two thousand reds for a season, and sometimes more. As soon as we arrived at the Shack we would prepare the slurry of red clay and water (as described earlier), dip the roots of each bundle of pines in the clay to protect them, and dig a short ditch "to spud them in" (as Dad called it). The ditch was in the shade west of the Shack so the pines seedlings would not dry out.

During the drive up our car was usually jam-packed with gear, and Gus or Flicky the dog. To keep things organized, we used the old chuck boxes Dad had used to lash to his packhorse when he worked in New Mexico. We generally stopped in Baraboo for a twenty-five-pound block of ice so we could keep our vittles cool. If Starker joined us he brought his little roadster to help carry the gear. We also looked forward to the guests sometimes invited to help us plant. Daddy's sister, Marie Leopold Lord of Burlington, Iowa, fit right in. She was lots of fun, and a great botanist with a special interest in ferns. One year our visitor was a forester Dad had met in Germany, Adelbert Ebner, who was a jolly fellow perhaps fifty years of age, and quite a musician. When Adelbert stayed at our house in Madison he sat at Mother's piano and began to play some wonderful German pieces, and he sang along with a strong, good baritone voice. He taught us some German folk songs, and his favorite composer was Franz Liszt. Adelbert stayed with us at the Shack all week to help plant. There were other visitors from abroad over the years, including (at different times) two famous British ecologists, Charles Elton and Fraser Darling.

Between planting sessions, we would take walks to look over the land. We used to go to Tom Coleman's hill above Lake Chapman and look for the early blooms of the pasqueflower. One time we walked all the way to the quaking sphagnum bog, nearly three miles downstream from the Shack.

Aunt Marie Lord (Dad's sister) helping Dad plant pines on the margin of the old cornfield. Camera faces southeastward over the future prairie.

Each year, Dad selected an area where we would plant the pines, and generally it was along the periphery of our prairies and along top of the Clay Hill top near Lake Chapman. Dad always reminded us to take good care of our tools, like polishing our shovels after use and keeping the tools sharp and rust-free. One of the important tools was the two-man crosscut saw, which was our chief way to make firewood for our Shack.

Each year, we began to plant immediately, after sharpening the shovels and forming teams. At lunch we returned to the

Tools. As a craftsman, Dad was always telling us that we need to respect our tools, keep them shiny and sharp, and not let them rust. Put them safely away. We did learn that.

Shack, warmed up, and made sandwiches to go with our soup. Dad often took a short nap after lunch so he could keep limber. That gave everyone a break.

In the evening of course we were busy getting ready for supper, building the fire up so we had good coals, and anticipating

Mother with our two-man saw, a critical tool for cutting firewood.

having a little *traguito* before supper. That was usually a time for singing and playing.

Poco and Pedro

During the spring of 1942, when I was in tenth grade, the university had its spring break a few days before we at the high school had ours. Mother and Dad, Nina, Carl, and the dog Gus had driven up to the Shack for a planting weekend, leaving me behind until my school break started. The plan was for me then to take the train up to Baraboo, and Dad would drive in and pick me up.

I had managed to find and raise a new crow. We called this one Pedro. His name was Pedro Gutierrez Montoya de Luna. Our house on Van Hise Avenue in Madison was right across the street from West High School, and behind our house lived Billy Haley, who was raising pigeons. The pigeons in Billy's garage made a constant cooing noise talking to each other. This meant that as Pedro was growing up he would make cooing noises much like the pigeons next door. He was a pretty good ventriloquist. He was also a friendly, courteous, and intelligent crow, who liked the pigeons. When the family drove to the Shack for spring break planting, Mother kindly put Pedro the crow and his box in the car. She promised to feed him till I got there.

After everyone had left for spring planting, I had a call from Dad's secretary, telling me he had called from the farmer's house to say that the river was now in full, flood. It was spring flood, as the river was up three feet, and had been as much as six feet! The river road was underwater, and the car was trapped at the Shack. So if I took the train from Madison to Baraboo, I would have to bring my bike and get in from Baraboo on my own. I doubt anyone was panicking. As Dad wrote in "Come High Water," "I see our road dipping gently into the waters, and I conclude (with inner glee but exterior detachment) that the question of traffic, in or out, is for this day at least, debatable only among carp."[1] Well, that is probably how the family was feeling up there at the Shack surrounded by water: delighted but stranded.

On Friday, the day before I was to leave, the postman came to the door in the morning and said he had found a tiny baby squirrel that had fallen out of his nest. Did I know anyone who could take care of him? I said, "Oh, let me see it!" The postman reached in his pocket and handed me this little furry fluff ball, just a youngster. He had light brown fur with specks on his back, and on his tummy was a delicate soft orange fur. He was so beautiful. I said I would be glad to take him in. The postman was relieved, saying that he had been worried about the little fellow.

I warmed some milk and with a dropper fed the little squirrel. He drank eagerly from the dropper, making soft noises. I came home at noon to eat lunch and feed the squirrel again. I decided to name him Poco, a good name for someone so small.

The next day I packed Poco and a few things in a knitting bag, put air in my tires, and biked down to the train station. Poco and I checked my bike and boarded the train for Baraboo. About an hour later I collected my bike and began pedaling northward toward the Shack, roughly ten miles along County Trunk Highway A. I took the entrance to Tom Coleman's property (which was high ground) so I could approach the Shack from the south, around Lake Chapman. That way I could bike over dry roads to the closest point. I left my bike at the Lake Chapman bridge at the road end and walked eastward around the lake, my plan being to cross the irrigation ditch and then head north to the Shack. The irrigation ditch was much wider than normal, deep, and well underwater. I held the knitting bag with little Poco high over my head and swam across the ditch. I was wading across the marsh toward the Shack when Dad and Nina spotted me, as I had on a red hat. They were up on the Sand Hill, planting trees. The way Nina tells it, they spotted something coming across the marsh but could not figure out what it was. According to Nina Dad began to laugh when he saw me emerging from out of the water, carrying a little knitting bag.

It was a happy reunion. I hugged Mother and Dad and showed them my precious new pet, the baby squirrel Poco. Here was something else for the family, especially Gus the pointer, to get used to—a squirrel. Mother was tactful. "That is very nice, Estella, but are you sure you can take care of all these pets, the Pedro bird *and* now Poco the squirrel?" Of course I thought I could do that just fine. Mother pointed out that we would be taking the train back to Madison. While she would carry Poco, she was not going to deal with the crow on the train. That meant that I would have to take Pedro on my bike. How was I going to do that? I assured her that I would figure it out.

We planted trees together all week on the edge of the cornfield, planting two-year-old baby pines following Dad's plan. As always, we worked in pairs, one person carrying the pail of trees, the other a shovel for digging the hole. We would get the tree in the hole, spread the roots, replace the earth and tamp the soil around the tree. Sometimes we would say, copying Mother, "OK, little tree, you *grow* now!" Someone asked me to pose and took a photo of Poco on my hat.

When the end of the week arrived, the road was still flooded, so the family walked and waded up the road out to Mr. Lewis's farm to ask for a ride to Baraboo so that they could catch a train back home. I said goodbye to everyone at the Shack and went back over the irrigation ditch, this time carrying Pedro instead of Poco. The water was lower now, but it was still a wet passage crossing that ditch. I got on my bike and wrapped

Poco the pet fox squirrel is posed on Estella Jr.'s head while we are preparing to go plant some pines.

some cloth tightly around the steering bar with string to make as a solid perch. I tied a shoestring around one of Pedro's legs to tether him to the bike handlebar. I talked to Pedro, asking him to behave and not try to take off. Amazingly, being young he did not even try.

I rode slowly along the County Trunk T, which at that time was gravel, and then down the blacktop of Trunk A. Pedro seemed to enjoy the ride, perched on the bike handlebar, facing forward and sometimes spreading his wings as we went along. He did not jump off the bike, thank heavens. We went trundling along fine until we were climbing the final hill approaching Baraboo on Trunk A. All of a sudden a pickup truck full of Leopolds went roaring by with the family standing in the back, all waving at me and loudly calling out "*Hellloooo*" as they passed. Off they went in a flash. It was such a shock that I stood up too quickly on my bike pedals and somehow tore the seat of my blue wool pants. I pulled into the cemetery above Baraboo to patch my pants with a safety pin. Pedro obliged, waiting. I then remounted, and we started down the steep hill into town. Pedro enjoyed the speed, crouching low on the handlebars and spreading out his wings as we went. It was great for both of us. He was quite a cyclist.

When we got to the Baraboo railroad station, Dad was sitting in the empty waiting room, reading a paper. He had Gus on a leash. He was waiting for me to appear, as the rest of the family had boarded an earlier train to Madison. We waited and waited for the next train. The station was empty. Dad read the paper. I finally went into the ladies' washroom with Pedro and tied him to the hot water tap, then went back.

After a long while, some women came into the station and talked with the stationmaster. I was sleepy and not paying much attention, but pretty soon they entered the washroom. Suddenly, they came rushing out, shouting to the stationmaster, "Oh, There's a raven in there! There's a raven in there!" Coming quickly to my senses, I went in and rescued Pedro, brought him back to our seat, and folded a newspaper around him, which he did not mind. He liked to cuddle. When the train finally came, Dad checked the dog and the bike. The three of us sat down in the train together. I held Pedro close to me, shielded with the newspaper.

The trip back to Madison was about an hour and a half. Pretty soon Pedro got hungry and began to talk a little. I pulled out some hamburger and gave him a bite or two. The sounds he made (a rising *caw caw caw*, then *gobble, gobble, gobble*) made the people in the railroad car look at me strangely. Dad buried his face in the newspaper. He did not want to be closely associated with this noisy bird and me just then.

We arrived at the Madison depot in late afternoon. Dad picked up the dog and the bike from the boxcar. The four of us looked for a cab. We walked around the platform, trying to find a cab that would take us all—the dog, the bike, the crow, and me and Dad. The first cab man looked us over and said, "No, you won't fit." We tried another cab but got the same answer. I'm not sure if it was Pedro or Gus or the bicycle, but we finally found one fellow who was amused at the entourage. He got us happily home.

And it was good to be home. Each Sunday evening after the Shack interludes, we would sit together in the front room at 2222 Van Hise, and have a soft-boiled egg and toast for a light supper. We would talk about the work we had done. Sometimes we would put some "good music" on the Victrola and listen after supper. Dad would sit and listen with closed eyes. Usually it was Brahms's Violin Concerto in D or Smetana's *Die Moldau*, which were some of the few classical records we had. Then Mother would read to Dad aloud from one of their exploration books. He loved to read about explorers in the Arctic or in the West, such as John C. Fremont's journals. One book they particularly loved was *The Way West*, by A. B. Guthrie, a novel about the mountain men in Wyoming, a good story colorfully told. As Mother would get sleepy, Dad would reach

Mother with Pedro the crow in the backyard at Madison.

over with a smile, pat her hand, and say, "Well, darling, I think it is time for us to go to bed." And so it was.

Poco and Pedro got along fine with Gus. In the evening Pedro would be put in his cage in the garage for the night, but Poco was free to explore. He often slept near my pillow, waking up early in the morning, anxious for more food. Pedro would do his usual thing, flying around the neighborhood all day. He would watch me and the other students go to school, which was across the street from our house. After school was out for the summer, Pedro visited the neighborhood houses, where there were children to play with and free food. When the family would have lunch out in the garden, Pedro was sure to be there, parading up and down the grassy lawn. He was watching for some morsel of food, of course. Poco was in the yard with us, learning to climb trees. He was very mobile.

When we all went to the Shack for the weekend, which was nearly every weekend, my pets would be brought along, as well as Gus. Poco and Pedro were set free to explore outside. Once, after cleaning up and washing dishes, we caught Poco trying to get his nose in the sourdough pan.

By summer, Poco had made himself at home inside as well—whenever I was allowed to bring him inside, that is. In our home in Madison, he found his way up to the mantel over the fireplace and would curl up in a Maria bowl, an extremely valuable black Navajo bowl from Santa Fe. We kept stick matches for the fireplace in it. Poco found that to be a good place to sleep.

When it came time for the elm trees to set fruit, there were hundreds of little wafer fruits hanging off them, sifting down onto the lawn. Poco climbed high in the trees. Dad took me

Poco the pet fox squirrel is investigating the sourdough pan on the table, but not for long, as we rescued him. Photo looks toward the south wall of the Shack; note the horn hanging on a peg in the background. Also see wooden buckets along that wall.

aside. "Baby, your Poco seems to like to eat the elm fruits. This is an excellent time to set the squirrel free, so he can become a wild animal and learn to sustain himself. What do you think?"

I can remember resisting this idea, as I was so fond of Poco. He was such a cute little fellow.

One time when I was not paying attention, Poco climbed up on Mother's walnut dining room table and chewed a three-inch section out of the hinged area where a folding leaf was let down, exposing a nice ridge of walnut. I tried to explain that perhaps he was strengthening his teeth, but I was in serious trouble with Mother. From then on, I had to watch Poco every second, and he was not allowed in the house again.

Later that month I turned Poco loose at the Shack, just as Dad had suggested. It was time, and the Shack seemed the best place. He seemed very self-sufficient there and enjoyed complete freedom to roam the woods. I am not sure that we spotted him again.

Sky Dance

April is the time when the spring peepers call. Pasqueflower (*Anemone patens*) and bloodroot (*Sanguinaria canadensis*) bloom. It is also the month when the woodcock present their courtship protocol. We children would eat an early supper and walk upriver at twilight to the best woodcock peenting grounds, a sandy prairie on a high terrace with little oaks we now call "Suevanna." When we got there, we took positions behind small oak trees and lay on our stomachs, keeping very still as dark descended. What a thrill when the male woodcock would come suddenly in with a flutter of wings and alight a short distance in front of us. Immediately he would give out a short *peent* call that we all thought resembled the turning of a cork in a wine bottle, with little hiccup sounds between the peents. It was thrilling to be only five yards or so from him, lying invisibly on the ground in the dark.

Then suddenly the woodcock would lift off and fly in circles in his sky dance, climbing higher and higher in the sky, making a kind of trilling or warbling sound that was nearly continuous. Then, just as suddenly, the sound would stop and he would drop straight down like a shot, brace his fall just before hitting the ground, and alight gracefully at the spot where he had been.

We found that when he left the ground and was twittering high up in his sky dance, we could jump up and move closer to the spot where he would land. We held our breath as he plummeted out of the sky. We think he never saw us, though we were at times five to seven feet away from him in the dark. He must have been intent on other things. It was a wondrous and exciting experience.

There were times we could stand in the front yard of the Shack and count the number of peenting males by ear. My parents called this "taking a census." For example, we might hear one or two on the downstream side of the cabin, and one or two toward the river on the island—perhaps seven in all. The woodcock courtship lasted for several weeks during April and May. The habitat they chose for their peenting grounds was open sandy savanna with smallish trees set far apart and sparse groundcover. However, we found that this was not where they nested. We happened upon a woodcock nest in grassy vegetation between some small oaks far from the peenting grounds.

We used to wonder where that woodcock's mate was, as we never heard or saw her. She must have been demure and cautious, never letting us see her, though once we accidentally flushed her off her nest by the Shack gate. Woodcock populations in Wisconsin are plummeting now. One cannot help but wonder if it could be because so much of our area of open vegetation has filled in with grass and become thicker, and we no longer have these great floods that deposit open sand on the lowlands.

Warbler Watching

In spring Carl, Nina and I would take binoculars and walk down into the Shack woods by the slough, hoping to witness the warbler migration. "Come on, let's go see what we can find!" Carl would say. Nina would chime in, "You bet! I'll get the bird book." In March and April these wonderful birds would come through our woods.

We walked up over the Sand Hill on the old cow path past the large glacial boulder we now call the Meat Rock (more on this later). We followed the trail down the steep side of the Sand Hill into the woods. Hooded mergansers were often on the Slough. Best to have your boots on, as there were many wet places in our woods. That glacial boulder marked the trail well. Somewhere around six feet in diameter, it is perched on the very top of the Sand Hill.

As soon as we entered the woods we could hear the warbler migrants. First, you could hear their songs; the most obvious one was that of the common yellowthroat singing *Wichita Wichita Wichita*! We would freeze and start searching for the bird. This was a time when the leaves of the soft maples were just beginning to open, so visibility was good. We would chatter at each other in delight at finding this warbler fellow. Walking on a little farther, Carl (who was a terrific observer) would spot another. Perhaps a black-and-white warbler. "Oh, handsome!" he would exclaim. "Get the book out and let's look at the picture!" Nina would say. "Yes, it checks out."

We knew that both of these warblers had flown at least all the way from the Gulf Coast up through the Mississippi valley to get to our woods. Imagine that they wanted to come visit *our* property on their way north. How wonderful! We wondered where they were going next. Probably further up north, as in the Midwest the black-and-white warbler breeds in an area north of us. "How in the world did these birds pass through Chicago?" we wondered. Oh, we heard that it had to be by flying high at night! Just think of that. Guided by stars. Years later, the son of one of Dad's friends and neighbors, Professor John Emlen, would answer that very question. Young John was using experimental means in a planetarium with caged birds. He found these migrant birds actually used the North Star. Pretty soon, if we were lucky, we would hear a woodland thrush, with its beautiful flute-like song. We guessed that these birds liked to hang out in the wet places looking for insects.

In addition to the first arrival of migrant bird species, Dad was keeping track of the blooming dates of our local plants, so we were on the watch for new first-bloom records for him. This effort is called "phenology." The date of first bloom is probably a response to the warming of spring weather. So a phenology calendar is a kind of biological record of increasing warmth during the spring season. Right away in March, first off in the wetlands we could see the skunk cabbage (*Symplocarpus foetidus*) with its big yellow spathe and spadix. Marsh marigold (*Caltha palustris*) with its flaming yellow flowers was in bloom usually by the twenty-sixth of April.

I recall a story about Dad when he was a student, taking long springtime exploratory walks in the woods around Lawrenceville, New Jersey. In one of his letters home he told his parents he could not see what the phoebe (which is an insectivore) could eat when it arrived so early in the Lawrenceville woods in March. At first, he said there were no bugs out that early in March in New Jersey. Then he trumpeted a new answer back to his parents: The phoebe was hanging around the smelly skunk cabbage and catching the tiny flies that were swarming around that early-blooming plant.

And so our journey went on. We walked upstream to Anchor's Woods, where we saw many young pine trees coming up between the giant white pines in that stand. We heard and saw the huge pileated woodpeckers, and we welcomed them. They are actually winter resident birds of mature forest.

When we returned to the Shack, we reported our findings to Dad and Mother, who were sitting in front of a nice warm oak fire, a welcome sight. Dad would ask what we saw and record the information in the Shack journal. Mother would greet us cheerfully. Something good for supper was simmering over oak coals on the fireplace. Soon it would be time to light the kerosene lamps on the mantelpiece, get out the guitar, pour a drink for Mother and Dad, and talk about the day's adventures.

Meat Rock and Calling to the Owls

In the evenings after supper we used to like to call to the owls before turning in. There was a very special place for us to do this—at what we now call the Meat Rock.

On the hill above the Shack the old cow path crossed the hill next to the Meat Rock. This convenient east-west trail was a remnant path undoubtedly made by the farmer's livestock. The deer used it too. At the pinnacle of the slope, one looked northward straight down toward the slough. At this very point stands the Meat Rock, a great granitic boulder, perched just where the glacier left it at the top of the hill. As Dad said, "This huge chunk of granite perhaps came all the way from Greenland." The boulder's south-facing flank is buried in the hill. The Meat Rock is immense, perhaps five to six feet in diameter, and with no visible scratches to indicate what tumbling it had been through in its long ride southward from northern Wisconsin (or even Canada or Greenland!) under or in the last glacier. It is mottled gray with little white specks of crystals buried in the rock.

We always admired this huge rock, and I can remember that Carl planted a little cliffbrake (*Pellaea*) rock fern just under the north-facing edge of the boulder, where it had a protected place to grow and thrive.

When we were crossing the hill it was always a temptation to walk around the south side and step up onto the huge boulder and look up and down the slough. It gave one a feeling of power to stand there, perched atop this great heavy landmark and survey one's surroundings. It was also a good place to stand and look at the stars without any branches interfering. Sometimes we would stand on the great rock in the dark. We would spot the Big Dipper and find the Cygnus cross and Orion's belt, the brightest constellations in the night sky. While we admired the great heavens we felt small and aware of the immensity of the universe.

One opportunity before bedtime was to climb up the hill and stand on this rock in the dark. In the spring, when the owls were calling, it was Carl's particular delight to stand there, cup his hands around his mouth, and call loudly to the barred owls. This was a descending call like this: *WOO woo*

woo wuh-WHOO. Then there was usually a follow-up call repeating this general sequence. The last note is a descending call with a gargle at the end. In the spring night, almost always, another barred owl would speak up and call back to Carl. Apparently barred owls (so named for the horizontal bars across their breast) are exceptionally territorial and want to investigate any owls trespassing on their territory, so the conversation with a resident owl would go on and on. Sometimes the barred owl would fly much closer to try and find out who this intruder might be. That was exciting.

Very rarely we could hear a little screech owl at the Shack. Their call is a descending warm whistle that has a warbling sound. We would all practice that call, which was difficult to make. It was in Madison particularly that one could hear the screech owls calling in the spring.

Once in a while, however, it would be a great horned owl that would answer. His was a softer call, lower in pitch, and something like *Whoo, wu hoo hoo*, with a pause after the first note. Their call was not as much fun as that of the barred owls, who were strident and anxious-sounding. The great horned owl's call, by contrast, was subdued, sounding less enthusiastic. Nina and I would practice over and over imitating Carl and the sounds of the barred and great horned owls.

After a while we would feel chilled standing on that rock and not moving much and wander back down to the Shack, put another log on the fire, and get warmed up again.

The Meat Rock got its name after *Sand County Almanac* was published in 1949. Mother had a little plaque cast in bronze that read "It is here that we seek and still find our meat from God" and had this attached to the great boulder. The phrase is from Dad's book introduction to *Sand County Almanac*.² Indeed, for us, that phrase has always been true.

Goose Music

In the old days Canada geese didn't stay put. Winters were cold and there was no green grass to keep them fed, so they migrated. That meant that they came back north to us in the spring. One time Dad had been in touch by phone with Frankie Bellrose, of the Cairo, Illinois, office of the Fish and Wildlife Service. While they were talking Frankie mentioned that thousands of geese that had been resting (or wintering) at Cairo in the marshes had just taken off to start their migration northward. Dad and Frankie looked at their watches—it was about nine in the morning. "Well, you can start expecting that flock of geese about two o'clock today," said Frankie. And sure enough, Dad's office could hear them flying over Madison at about two that afternoon, calling and honking. Dad commented, "That means they flew at about fifty miles an hour!" How wonderful!

It was a particular treat for all of us when a horde of Canada geese would come through and land in our marsh. One could hear them all evening long, honking at each other, and even making a few honks at night, perhaps the "border patrols" talking to one another. Dad and Mother took special delight in listening to the geese in the marsh. Goose music, he called it.

One time late April Dad and Mother and I carried a bench and a stool out into the Shack marsh during the day, when the geese were busy visiting a nearby cornfield looking for something to eat. This furniture would be our seats in the goose theater that evening. So at sundown we wandered over and sat in the marsh to await their return. It was indeed just like a theater where one awaits the curtain rising. Our actors, the geese, had already started flocking into the marsh. In they came, usually in small groups, perhaps two or three families of geese flying together from Mr. Lewis's (or someone's) cornfield into our marsh, getting ready to spend the night. Flock by flock arrived, swishing the air with their wings and calling to each other. They landed with a plunk and kept right on talking. Dad took particular pleasure in watching all this, all the while keeping track of the size of the flocks for the Shack journal. He entered a record of 642 geese on April 11, 1946.[3] The average size of the flock was six geese, which, he suggested, might be the number in a single family.[4]

He was so tickled to hear all this goose music. He and Mother would giggle and say, "Oh, listen to that one! They are so excited!" The most important part of all this to us was that the geese had returned. They flew all the way up the Mississippi River and landed in *our* marsh.

What Species Do the Deer Prefer?

During certain springs Dad would invite some of his graduate students to come visit the Shack area with us. The point was to get them to learn to be observant in the field. The professor would walk them down among the plantings and explain the kind of problems we have had getting the small pines through the season. He walked using his cane, acquired on a trip to Germany, as a pointer and asked such questions as, "What kind of browse is this?" He showed the students our growing prairie, and what we were doing to introduce new species to the old cornfield. Then off they would go to the woods, where one could look for signs of browse on the low shrubs. "What species did the deer prefer?" Dad would ask, just as he did with us. What happens to these plants when the deer population is high?

Then it was off to visit the wetland at the east side of the Sand Hill. Dad explained we were trying to introduce tamarack seedlings here. He was not sure why they were not doing well when it seemed that all their needs for habitat were met.

At certain times in the spring Dad would actually stage an undergraduate class field trip at the Shack. This was for Wildlife Ecology 118, a three-credit course. I remember one story told us by Charlie Bradley, who reported that most of the class seemed to be bird watchers or even ornithologists. The class

Photo of Dad and his class on a field trip in winter.

was walking down on the sandy shore of the Wisconsin River when Dad turned to the class: "Heads up, you all. What's that bird sitting by the water's edge?" He pointed with his cane. There was a great silence, perhaps as people were trying to run through their knowledge of birds. So geologist Charlie spoke up cautiously: "Sandpiper?" The class people apparently turned on Charlie and replied to him with vigor: "Yes, but what *kind* of sandpiper?" Charlie said he was appropriately silenced by that one. He always laughed hilariously when he told that story.

After Kumlien Club (a kind of seminar) meetings at Dad's office, the grad students, wonderful people and hard workers, often came over to our house in Madison and sat together in the living room, talking and telling stories, or discussing some issue. Dad and Mother usually served apples or cider. When I was little and upstairs in my pajamas I would sometimes sit at the head of the stairs, listening. One day the students got to singing songs they made up. The first verse was made by Albert Hochbaum, an artist and a graduate student of Dad's. It went like this:

> The rabbit has a funny face
> Its private life is its disgrace,
> You'd be s'prized if you but knew
> The *awful things* that rabbits do.

They all got laughing then. And Fran Hamerstrom or someone came up with this verse:

> The skunk he leads a lonely life
> With just his children and his wife
> And though he has a social bent
> No*body* likes him for a scent.

Someone else came up with this:

> The horned owl nests in ice and snow
> When temperatures are ten below
> I should not think it very nice
> To *incubate* a cake of ice!

They were having a great time. Then someone coaxed Albert to sing a song they knew he liked. Everyone laughed and laughed to see this great big man singing this song at high speed.

> My gal Irene, she's the village queen
> She's the queen of the village green
> When she plays on her accordeen
> Prettiest thing you ever seen!

On another occasion the same crew was gathered in the living room after Kumlien Club and talking about owls. I stuck my head outside the storm window upstairs in my room and tried to whistle like a screech owl. I repeated it several times, when all of a sudden I could hear Fran's voice calling out, "Oh listen everyone, there's a screech owl out there. Right here in the city!" And they all jumped up and ran to throw open the front door and listen for the little owl. I thought that was LOTS of fun!

Road Kill for Supper

Fran and Hammy Hamerstrom were two of Dad's graduate students who we all got to know fairly well. Fran (Frances) was from the East Coast and had an erudite accent; Hammy was probably from the East too, but his accent was not as pronounced. Fran was a lively enthusiastic person, Hammy more soft-spoken, pensive;

both were handsome, interesting people. While we were starting our work at the Shack they occasionally came along to help.

The Hamerstroms were both expert ornithologists, and as a part of their graduate research they were studying Wisconsin's native prairie chickens. To do that, they found and occupied an old farmhouse in Plainfield, Wisconsin, on a property near where the prairie chickens would carry out their mating and nesting rituals. One weekend in the spring when our family was all at the Shack, the Hamerstroms came by to visit and began talking excitedly about watching the prairie chickens dance on their booming grounds at five in the morning. Nina, Carl, and I were fascinated with their description of it, and asked a number of questions. Fran, who had a kind of distinctive breathy voice, said in her enthusiastic way, "Oh, we are going up there right now, why don't you come along and we will *show* you the prairie chickens dancing in the morning!"

We said, "Oh, we don't want to be any trouble," and Fran said right back, "Oh, goodness no, you will be no trouble at all, and we have room at the farmhouse. So come." Thereupon we piled into the Hamerstroms' backseat and off we went. All the way there Fran and Hammy were talking about their work with these birds and how interesting they were. It was a good two hour's drive to Plainfield. On the way there we passed a farmhouse, and a chicken was lying on the road, apparently hit by another car.

"Stop the car!" shouted Fran.

Hammy stomped on the brakes.

"Carl," said Fran, "Would you kindly run out and pick up that chicken for us? We need it."

Carl, surprised, obediently stepped out of the car and picked up the chicken. He got back in and slammed the door.

"Now," said Fran, "if you would, just open the window and begin plucking all the feathers off of the chicken. You, Estella and Nina, can help." She added enthusiastically, "We cannot let that chicken go to waste!"

So we dutifully began plucking the feathers and throwing them out the window. The feathers were fluffy and some got in your nose. It was quite a big job. We stayed at it until we had that bird pretty well plucked.

"We are going to have this bird for supper tonight!" Fran announced. Nina and I raised our eyebrows, and said, "Oh! We are?"

This event reminded me of one of Robert Pyle's stories about road kill; he quotes John McPhee's essay "Travels in Georgia" "about the pleasures of road-kill cookery." Pyle said that he finds that "D.O.R. (dead-on-road) animals, if fresh and intact, furnish elegant opportunities for highly instructive dissection."[5] Well, that is another view of road kill.

When we got to the farm, Fran had us help build a fire and pump some water. At least we were used to doing that. She proceeded to cut up the bird and cook it.

While we were awaiting the meal she took us out to a small shed and introduced us to her pet harrier hawk. It was a huge bird, and apparently very used to her. Fran wrote a book about this bird later on; he was her special pet.[6]

We were surprised that the chicken tasted pretty good. Yes, we did get up very early the next day and watch the prairie chickens dance in the early morning light. Most of all, however, we all remembered the road-kill chicken.

Four

Summer

Summer was a time for transplanting prairie wildflowers. We knew that we wanted to restore prairie on the cornfield in front of the Shack. How did we know where we could get these prairie species? Of course there were no commercial sources at all.

We had heard that prairie species were especially prolific along railroad tracks, because in those days the railroad frequently burned them to control brush. So we would stop there during different parts of the summer and find the prairie species in bloom (so we could identify them), or along an old road cut where we felt we could dig up chunks of sod with the species, put them in a tub in the car, and transport these to the Shack, to spud them in to the old corn field (our future prairie). This included prairie grasses, legumes, asters, and a whole variety of perennial species. And of course these can reproduce. This means that in those days (and to some extent now) there were "idle spots" along each side of the railroad tracks, as Dad observed, where the cow, plow, and mower are absent and a profusion of wild prairie herbs persist and bloom vigorously.

Some species had huge deep roots, like the beautiful compass plant. Dad collected their seeds and built a little plot on the hill to plant these along with a mix of seeds of prairie grasses. This was an experiment. As mentioned, he did not water them, but they came up and did beautifully. So we knew how to promote such species on our prairie. (See chapter 7.)

Over the years our prairie became more diverse, and more beautiful. According to the Land Institute of Salinas, Kansas, these native perennial prairie herb species typically grow very deep roots. Some extend downward ten to eighteen feet below the land surface! So it is no wonder the prairie vegetation is so stable and tenacious during drought; they have unusual adaptations to reach moisture and minerals at depth.

Silphium plant in bloom at Prairie Birthday site, Prairie du Sac. These plants are five feet tall!

The Rhythms of Summer

Summer at the Shack was an idyllic time. The family ritual started with Dad getting up very quietly, sometimes as early as 4:00 a.m., or even 3:00 a.m. when he was checking on bird songs and light.[1]

Dad would build a fire in the yard fireplace and make coffee out there, listening to the early birds with his light meter. He was measuring how much light there was as each species began to sing in the morning chorus. He was of course recording the data in his notebook. As the sun warmed the air, he went on his morning walk with Gus.

When the sun came up, Mother would rise, and then us children. It was always so pleasant to step out barefoot onto the dewy grass and walk to the Parthenon, studying the pretty birdsfoot violets (*Viola pedata*) blooming along the path. In Dad's prairie garden in front of the Shack, we would check out the gorgeous spiderwort (*Tradescantia ohiensis*), which you could always count on producing one new fresh blue flower with three petals and a yellow center every single morning all summer.

Upon gathering around the fire, the next step was to add things to the sourdough for breakfast pancakes. We always mixed the yeast, flour, and water the night before. In the morning we would add one or two eggs, some salt, and definitely sugar, then beat the dough and set it to warm near the fire.

When the sourdough started releasing bubbles was the time to start cooking the bacon in the Dutch oven, and then the pancakes. About this time Dad and Gus would be arriving back in camp after their walk, often with stories of what they'd seen. Dad would announce things like, "The red osier dogwood is in bloom," or maybe, "The towhee has arrived."

He would then start cooking pancakes, usually using bacon in the Dutch for each cake—to spread the grease over the bottom of the oven. It was important to Dad to have the plates warming on the trivet by the fire, so that when the cake was done he could put it right on the hot plate and hand it to one of us. He did not like to serve cooked food on cold plates.

We loved eating cakes with butter and syrup. A special treat was to have some of Ed Ochsner's honey with honeycombs of beeswax and then roll it up and cut off sections for mouthfuls of pleasure. Oh, how we loved those cakes. Even when we ran out of cooked bacon, we kept eating cakes till we were full. Dad was usually the last one to stop cooking and sit down to eat. We also found that the cakes were good cold, for making sandwiches for lunch. So on Sunday morning, we used up all the batter.

The trick for cooking sourdough pancakes is to have plenty of oak coals. The fire has to be fed ahead of time so the coals are there. When we cooked inside, Dad would take a tiny shovel and reach between the andirons to scoop out the coals he wanted. He would set the Dutch oven right on those coals to get hot and repeat the process as the coals lost their heat. Of course the good, dry oak made the very best coals. Pine, maple, or birch cannot hold a candle to oak coals.

By this time we also had the teakettle boiling on the fire, usually hanging from a wire hook. We would be washing dishes on our great oak slab table in the yard. It had enough room on it for the big dish pan and a smaller rinse pan, as well as a place to pile the dishes. We often used Mother's wooden salad bowl for rinse water.

Washing dishes out side in summer. Luna is washing, and Nina and I are wiping, while Mother looks on. This particular table is a heavy, thick piece of oak with maple logs for legs. Mother is sitting on a "Leopold bench."

After the dishes were put away, it was time for Mother and Dad to take their walk together with Gus. They would stop along the way and admire the pines or whatever was blooming or fruiting—and record phenology items. On their walk together, Dad would show Mother the things he had noticed on his earlier walk. When they came back, Mother typically had a handful of fresh native flowers. She would set them in the bucket in front of the pump or in a vase on the table for decoration.

Dad is admiring his pines, now half grown. He said he was "in love" with pines.

After walks was time for projects, such as building furniture, fixing the bunk springs, working on the Shack roof, or doing whatever else needed doing.

Mother and Dad would call us to lunch by blowing on a special cow's horn Dad had hollowed out and then affixed a bugle mouthpiece to its narrow top end. We also used this trumpet-like horn to signal to each other when we were on hunting trips (note that the horn can be seen hanging on the wall behind Poco in the sourdough photo, along with the wooden buckets: chapter 3, figure 6). After lunch Dad and also Mother would take a nap, and we children would be very quiet, or go for a walk ourselves, or climb up to our tree house, about which more soon. This was a good time, too, for me to visit my island. We usually had a post-

Mother at the pump; she has just returned from a morning walk with Dad and brought her usual fresh flowers to decorate our table.

nap, pre-dinner project going—for example, repairing the slats on the outside of the Shack walls or building a new bench.

In the evening, Mother would get the supper ready to cook and a good fire going, activities in which we would take part. When the preparations were set, Mother and Dad would have whiskey sours in little glasses. They knocked a small chunk of ice off the twenty-five-pound ice block in the wooden ice chest, and added a shot of bourbon, water, and a twist of lemon. As dark approached, we would light the candles and the kerosene lamp. We would all sit around the fire and talk or perhaps sing a song or two with the guitar.

Washing dishes; Dad and Carl are working on dinner dishes.

Dad usually cooked the meat, often a game bird, in the hot Dutch oven. Mother cooked the vegetables or rice. Very tasty. We usually had homemade bread that Mother would have made in Madison. But on long weekends we would make and bake bread at the Shack. Using hot plates from the trivet again, Dad would serve, and we'd all sit around the table and enjoy the good supper. With everyone helping, washing dishes got done fast. We would then sit back down and play the guitar and sing. Someone would tell stories, and we'd generally make plans for the next day. Dad would usually turn in early, as he was a light sleeper and an early riser. The rest of us would sit around the fire and continue the singing and talking till our bed time. Carl or I would take turns playing the guitar, and we would all sing Spanish songs. We loved to harmonize on the Spanish folk songs Mother had taught us. She had a lovely voice and was very musical, having played the piano for many years. Sometimes, Nina liked to recall, after we all thought Dad was asleep, suddenly his voice would pipe up and he'd ask something like, "How about playing Brahms's lullaby?"

On Saturday night before we went to bed, Carl or Nina and I would get out the cube of Fleischmann's yeast and the bag of flour and go to prime the pump and get some water. We would stir the yeast and then the flour with water into the white sourdough pan, beating with the cedar paddle that Luna had made to get all the lumps out. It had to be a certain texture so it would pour well in the morning. And we would place the sourdough kettle, with its lid on, somewhere near the fire to keep it warm during the night.

To get into our bunks, we would stand on a bench, reach up, and grasp the Osage orange stave screwed to the ceiling, lift our legs, and pop into the bedroll. The original bunks spanned the entire wing of the Shack. With its straw covering, this was at that time a soft and fragrant bed. It was OK later when we had real mattresses too.

Tree House

During the early years of our time at the Shack Luna declared that we had to build a tree house. About twenty yards from the Shack, along the path up to the old foundation, there was a good-sized elm tree with big spreading branches and a good base for a tree house. Luna and Carl and Nina and I began to scout around looking for some real beams and boards to start construction. There was no way to climb that tree without using a ladder. From the ladder it was easy to climb up the rest of the tree. At a point, about forty or fifty feet above the ground, we began to drive spikes through a couple of beams (about 2″ x 10″) to pin these to the branches of the elm tree. This was quite an operation.

We began to pull thick boards up the tree using a long rope and a pulley, and soon could lay boards across the two beams to make a nice platform for sitting, and then nailed a board on the west side of the tree house to serve as a backrest. That made it secure and safe to sit with legs straight out leaning against the backrest. From there one could look down across the Shack yard and eastward toward the new prairie.

One of our occasional visitors was young Reed Coleman, whose father, Tom Coleman, had bought property on the far side of Lake Chapman. Reed once kidded me with his remembrance of arriving at the Shack with his parents on a Sunday. As they drove in, he saw me climbing quickly up the elm tree and staying put up in the tree house for a time. Well, that was kid stuff, of course, but it was a great place to haul up the guitar and sing and play songs together, or to practice, especially when Mother and Dad took their naps. Climbing up to the tree house was a favorite thing to do in the warm season.

There was one time when Nina and I were determined to learn the words to a certain Spanish song. It was "Naranja Dulce." We climbed up into the tree house with the songbook in my vest and a long rope to pull the guitar up after us. We were sitting there singing and worked on that song, when suddenly it began to rain. Then there was a big rush to cover the songbook, lower the guitar down to the ground, and climb down before either got very wet. Well, the songbook still has splash marks from that watery midday. The guitar survived the trip.

Leopold Benches

Dad and Mother loved having benches along the trails, in the yard, and elsewhere, places to sit patiently to watch a bird or to rest quietly together during a walk. The first bench was right in front of the Shack, made of a long, wide cottonwood board that curved beautifully to make a backrest. We built it against one of the old elm trees. The seat consisted of a long pair of beams, providing enough room for several to sit. We then built the table out of a very thick oaken beam for which Dad and Luna made four legs of maple. This required drilling with a large hand auger four holes at an angle in the bottom of the oak slab to insert the legs into. This was the table on which we typically would eat lunch and wash dishes. Dad and Mother would have us haul the heavy table into the Shack each time

The very first bench we built used an old cottonwood board, bleached by the sun and split, and curved to form the back. We attached it to an old elm tree in our front yard. Here I sit with my new guitar, ca. 1936. Sand Hill in background.

we closed up the cabin preparing to leave. I always doubted anyone would steal this enormous slab of wood, but we loved it, so we would move it in and out each weekend trip to keep it out of the weather.

After that Dad began to build portable benches, ones that could seat two and that would be easy to carry to any place he and Mother might like to sit. Once he had picked out the wood for the seat, he cut two long boards to go one way at the end of the seat and two shorter boards to go the other way, forming the legs of the bench. These were nailed in an X pattern at each end of the beam that was to be the seat, using braces, of course. Now what was needed was a backrest. After sawing the long boards forming two of the legs at a comfortable angle (calculating this involved getting Mother to sit on the bench seat leaning back a bit, and determining exactly what angle would be most comfortable for her), the backrest board could then be attached to the "legs." This pattern has now become known by friends as the "Leopold bench" design. It is just a practical quick way to build a useful bench. The design has caught on in recent years.

One spring, in our exploration of the patch of willows along the slough and my island, we ran across a two-by-four made of cottonwood. It was about ten feet in length and weathered. It had lain in the sun, ridden along the river, and dried thoroughly in the weather so that it warped nicely into a long curved J shape. As it was cottonwood, we knew it would be hard to drive nails through. We had discovered that when building the bunkhouse and using one or two river-borne cottonwood two-by-fours for roof beams. Many a nail got bent as it refused to go into such a beam. Sometimes we had sore thumbs from trying to hold the nail when it buckled. Ouch.

The one thing you can say about cottonwood two-by-fours is that they weather into the most artistic shapes, unlike pine and other construction beams. When Dad saw this particular two-by-four he knew exactly where we should use it to build a new bench. We built a great number of benches. Several of them were built into the landscape, and several were portable, so that we could carry and park them by a trail someplace.

While Dad and Mother casually built a practical bench like this, the portability and simple design has caught on. People call this design the Leopold bench.

This one was going to be a stationary bench attached to a couple of trees. We always called it our "Slide Hill bench." This particular site was at the crest of the hill above the Shack where one potentially could look over the slough and watch for mother ducks and their babies swimming about in the spring. The site was about thirty feet east of the Meat Rock. The bench site was on a slope between two middle-sized elm trees; these would serve as a perfect bastion for attaching the bench. The elm trees were about eight feet apart.

Nina and I were with Dad and Mother that particular summer weekend. Dad said, "Let's get some tools together and build us a bench right here!" We gathered up a crosscut saw, some heavy nails, a hammer, and an ax and carried two other planks we had found in the willows that would just span the distance between those two middle-sized elm trees on the slope. Mother was sitting nearby with her knitting and cheered us on.

The first step was to attach the two flat straps to the elm trees to hold the seat of the bench. We sawed the length of the plank seats to fit exactly between the two trees. That meant putting up a pine two-by-four support and nailing it to each elm tree as a horizontal crosspiece to support each end of the two planks. With an ax we cut a small groove, removing bark on each elm tree so the support struts had a flat place to nail to. We cut the length of the two planks and started to fit them to the space between the two elms. That was when we discovered that we were constantly sliding on our bottoms down the slope. It was very hard to get the crosspieces nailed properly, as the bluegrass on the steep incline there was slippery. Finally, we got these boards in place and began to attach the planks on top of them.

After a break for lunch—as this was getting complicated—we nailed the planks in place, which helped a good deal, as we could hang on to them (to avoid sliding). Then we stood behind the bench and fitted the curved two-by-four to the elm trees as a backrest. When it was all done we got out the camera. We took a photo of Dad sitting triumphantly on the new bench, and there is also one Nina took of Dad and me sitting on the new bench. It was Dad who named our construction the Slide Hill bench. And that is exactly what it was.

Benches. This was the "Slide Hill bench" that Nina, Dad, and I constructed one summer day. It overlooked the slough. It was attached to two elm trees and sported an old cottonwood board backrest. Dad and I were enjoying the view.

When we sat there we began to wonder what we could do about the pine trees Dad had asked Carl to plant on the steep slope in front of the Slide Hill bench. Clearly they were going to grow up and be in the way of the view of the Slough. Well, they grew slowly, and eventually the elm trees perished and the bench melted from age. But we certainly enjoyed the seat on the hill over the next few years.

Our Beach

Summer included trips to the beach along the shores of the Wisconsin River. It was a beautiful wide beach of pure sand. The family would take a break from our chores in hot weather and go down to the river for a swim. Some of the time we went bare-skinned, with the men swimming on the upstream side of the beach and Mother and Nina and I on the downstream side. If it was just the family, that seemed appropriate. Other times we all wore bathing suits, so that we could all play hop, skip, and jump. That was always fun, even though the big folks seemed to always win, jumping the farthest. Sometimes when Dad was feeling up to it he would do handstands on the beach for us. And he could even walk around on his hands. We were impressed! He had been an athlete when he was a student at the Lawrenceville School, and he remained fit all of his life.

In our early years at the Shack, Dad's brother Carl Leopold of Burlington, Iowa, president of the Leopold Desk Company, would drive up with his family to visit Dad and Mother. Uncle Carl's wife was Dolores Bergere Leopold, Mother's younger sister. So there was a double connection between our families.

Caryl Leopold, their youngest, was just my age. After their visit, Caryl would stay on and spend the summer going back and forth to the Shack with us. It was the beginning of what became an enduring friendship between us.

We would all drive up on Saturday mornings. Caryl and I would help unpack the car, prime the pump, and pump buckets of fresh water. Then, as soon as we could be excused, we would run down to the river, take all our clothes off, and swim. On that beautiful beach we would build castles in the sand, play hop, skip, and jump (competing, of course), and find clams in the shallows. At very special times we would walk along the shores to find where the river had "painted" green along the little pool margins. In these would grow the tiny plants (about two inches tall) on the shores that Dad described so well in his essay "The Green Pasture" about the river's "paintings on the sand": "I know a painting so evanescent that it is seldom viewed at all, except by some wandering deer. It is a river who wields the brush."[2] The tiny plants of *Eleocharis* make a delicate green border along the pool of water, and over the course of a week or so they grow into a dense bright green "garden" along the shore, the stems reaching a length of only about two to three inches. These green patches of moss-like plants (sedges) flourish in the calm margins of a lagoon and shoreline. Dad's essay deftly describes the river's ability to paint these borders of *Eleocharis*, and sometimes baby cottonwood and elm seedlings two inches tall would appear. These plants decorated the thin line along the shore, sometimes sparkling with dew. Caryl and I would spend endless hours down on the river sandbars, a beautiful place.

When interesting logs would float down the river we were there to catch them. We caught one memorable and very heavy log about ten inches in diameter and about four feet long and dragged it up on the sand. It had been roughly hand-hewn to make it squarish. It had a large oval hole at one end, probably to accept another beam to build a frame. We hauled this log upstream in the shallows, and then leapt onto it and rode downstream, hollering and happy. We did this over and over again. We named that log Napoleon and eventually convinced Dad and Mother to help haul it up to the Shack. At first Dad made it into a log seat next to the outdoor fireplace. Later he drilled four holes in its flat side and inserted four maple legs that he trimmed with a drawknife. Then we could sit on the Napoleon bench in the yard as we ate lunch. Eventually someone found a wide pine board to nail across Napoleon's back to make a coffee table (sometimes used as a bench). That long heavy table bench, old Napoleon, is still in our Shack, more than sixty years later.

The wonderful thing about our open sand on the river beach was the artistic pattern made by the small wisplike grasses growing there. There were little patches of a grass, perhaps a *Panicum*, a single plant six to eight inches tall standing alone in the sand, its stems reaching out like radii where the wind drew circles around the clump of basal leaves in the center. The wind blowing from different directions would gently move the grass stems as though they were spokes on a wheel, and the tips would draw little circles in the sand like a compass. Each plant then was surrounded by these artistic circles, making the plant like a beautiful target standing alone on the orange-brown sand. The doves loved the seeds of these grasses and were frequently there visiting the beach.

Grasses on our floodplain cut delicate circles in the sand as they are pushed by the wind. This is one of Carl's photos that won a prize at the university competition in Madison. A COOL photo!

The floodwaters of the spring would expand over the beach, depositing new sand that covered the old surface. As the water level dropped, the gentle wind and the moving water cut little tiny waves on the sand surface. These were like a marbled surface, and as the sun lowered it would cast shadows across each ripple. The beach was once huge, and these patterns cut by nature were beautiful and regular. One hated to walk on it for fear of disturbing something sacred.

Our beach and swimming hole. After a spring flood, the ripples preserved in the sand form a beautiful pattern. I am nine years old here with a small sailboat. One can see the Parthenon on the ridge behind me. Photo looks south. Carl won a prize for this photo at the University of Wisconsin.

On the Shores of Lake Chapman

During spring and fall floods, the Wisconsin River spreads out over its floodplain, and its waters flow along in channels through the riverine woods upstream from the Shack and southward and eastward over the great marsh.

One particular channel, which may once have been a main channel of the river, passes through the narrow Lake Chapman to the south. When the roads were laid out in the 1800s, one road intersected Lake Chapman. A bridge was constructed. By the 1930s that bridge was apparently declared unsafe and the floorboards taken up. But the vertical pilings still stood. For each strut of the old bridge support there were three pilings. When we first went there in the mid-1930s we loved to swim in Lake Chapman, as the river water was at that time brown-colored and with some clumps of paper pulp. Lake Chapman's water is deep and clear, not brown in color except during floods. We also loved to fish in Lake Chapman, sitting near or on the bridge pilings, and occasionally catch little

Luna with a nice walleyed pike he caught from our duck boat on the Wisconsin river.

The bridge supports at Lake Chapman. Carl is preparing to hoist himself up for a bit of fishing. Flicky the springer spaniel wants to go along. Photo looks toward the southeast.

bluegills, which were great for supper. In later years many other locals would come to fish at the site (some, unfortunately, leaving behind bottles and cans).

One early summer, Nina, Carl, and I were sitting on the bank opposite the first bridge piling when we noticed a very pretty warbler sneaking up and down and along the piling. The bird was breathtakingly beautiful. We watched it enter an opening that had formed just above the waterline. When it reemerged we began to realize it was a rare prothonotary warbler, the first one we had ever seen! Carl, Nina, and I sat, transfixed, watching this gorgeous yellow bird with its blue wings. We dared not move. The bird was apparently nesting in this cavity: it flitted back and forth between the piling and the bushes, where it may have been finding insects or whatever to eat. We sat quietly for a very long time, observing with glee this remarkable bird, a once-in-a-lifetime vision, we thought.

I did see a prothonotary warbler again in my life, but I shall always remember that marvelous first encounter, hearing it chirp and watching it move along the piling. In later years Nina, Carl, and I talked occasionally about the experience we shared on the shore of old Lake Chapman.

What We Found in the Sand Blow

One summer when Caryl Leopold was staying with us, she and I devoted many hours to building castles in the sand up on the sand blow, the ridge top—probably the top of a glacial moraine—just west of the Shack. This was an open area about twenty yards in diameter where no plants grew. Not one.

It was pure open yellow-orange sand, medium to coarse in texture. One could see the sand blow from the river road and from the Shack. There was always a good deal of discussion about why no plants grew on the sand blow. Some of the local people thought some farmer had dredged off the topsoil and put it somewhere; others believed an early farmer had tried to plant corn on that hill and turned it into a sandy desert.[3]

All we children knew was that it was a great, thick bed of sand for digging and for building castles. On one summer day, Caryl and I were digging deeply to make a "fort." Our excavation went down about three feet. At that depth it was still pure orange-yellow sand. We were still digging when we began to see little chips of colored chert. Not knowing what that meant, we kept going, and pretty soon we had dug up a beautiful little Indian arrowhead point of pink and white chert! The point was about two to three inches long, and it was broken. We triumphantly carried this down the hill to show to Mother and Dad. That led to a good deal of discussion at night about how this Indian point came to be buried so deep in the sand there.

Much later, when I was taking geology as a sophomore at the University of Wisconsin, I began to realize what this buried point might mean in terms of our local history. After we learned from the pollen analysis nearby about the past climate and ecological history of the Shack area—which had once been practically tundra, then spruce forest, then pine forest and hardwoods—we were able to put some ideas together.[4]

Imagine the time when the glacial ice to the east had abutted this single morainal ridge, which we now call the Sand Hill and Clay Hill (see Map 1). The climate would have been frigid, cold, and windy. This ridge may be a retreat moraine (a ridge formed by the local glacier when it stood for a long time in one place dropping sand and sediments at its terminus). The ridge lies perpendicular to the river channel. When it was formed, the periglacial winds blowing along the Wisconsin River bottom may have been very fierce. The winds perhaps blew this coarse sand up from the floodplain on top of the ridge, forming the sand blow, a very thick deposit close to and above the river channel. From the carbon-14 dating obtained by Louis Maher and Barbara Winkler in Lake Chapman, we know the ice was retreating from this area by about 12,500 radiocarbon years ago.[5]

In order for that Indian arrow point to have gotten buried three feet under all that pure massive sand, there are alternative possibilities. One might be that a prehistoric hunter or resident, perhaps some thousands of years ago, might have been there sitting in the sand trying to make a living in this frigid landscape. The chert chips tell us that he likely sat right here with sand blowing around him, chipping away. Alternatively, the burial of this small (four-inch) arrowhead point could have been during any particularly windy, dry time after the Ice Ages. Perhaps it was in mid-postglacial time—long before the first farmer came along in the 1800s to settle the area, and long before us. The Indian discarded the point, because it was broken. What an image in our history! Finding evidence of this simple act should give us an appreciation that we are not "special" residents in the area. Our species may have been on this land for a long time, sometime since the glaciers melted. When was that? Well, three feet of sand ago. That should give us real pause for thought! In any case, I really like to think about those scenarios.

Later Years: Building Trails

That was Shack life, and it was wonderful in retrospect. At first it was Luna, Carl, Nina, and I who were there with Mother and Dad. Starker had left to work temporarily at Coon Valley when we began our regular family outings to the Shack; then he went on to graduate school. When Luna took a job in New Mexico around 1937, that left three of us. When the United States entered World War II, both Carl and Luna were off to the military, while Starker had gotten funding to do a field survey of wildlife in Mexico; so that left just Nina and me. Then in 1939 when Nina got married, it was just Mother and Dad and me.

I was apprehensive about losing my dear sister companion when she got married. We went on a real canoe and camping trip, just the two of us, down from Neenah, Wisconsin, floating several days down to the Shack, before she married Bill Elder, which took her away. Throughout the war, with my brothers enlisted and Nina married, I was the only one at home with Mother and Dad. All through World War II and afterward, our trips to the Shack were as regular as clockwork.

One of our weekend projects was to build trails. In order to explore the woods, we had to have good trails. Building them was hard work, especially when it came to cutting through prickly-ash thickets (*Zanthoxylum americanum*). I vividly recall walking behind Dad, both of us with nippers, cutting prickly ash. We would walk the edge of the slough to the branch creek that led to the Bee Tree and the Otter Pool, as

Slide Hill bench; a rear view. I am wondering how primitive man made a living out here.

Dad called them, clearing the prickly ash as we went. Right along that branch creek was the densest thicket of prickly ash you ever saw. It was summer, and the mosquitoes attacked us in dense humming clouds. Dad seemed impervious and paid absolutely no attention to them. But me, I was frantically wiping them off my forehead and swatting them off my legs and pants. Even wearing gloves, I got bites on my hands. They seemed to go right through leather, canvas, everything. Dad just kept cutting and acting as though the bugs were not there. With Dad one did not complain, but it surely was miserable.

Later we cleared trails from Anchor's Woods leading downstream to Sand Hill. At one time these were excellent for cross-country skiing. One of these trails was later named after a dear friend of the family, Eddie Gordon of Madison, who perished in a snowstorm in the Rockies.

Five

Fall

Bounty from Our Shack Garden and Orchard

In the fall we had great fun picking our orchard apples and harvesting in the garden. There were two old apple trees, undoubtedly planted by the Baxter family in the late 1800s. One bore very large sweet apples, which probably was a Wolf River type, in Mother's estimation, and the other bore just nice, tasty apples. The trunks of these two trees were ten to twelve inches in diameter, so they were really mature trees. Under and around these apple trees we usually had planted potatoes (our best crop!), corn, and huge beefsteak tomatoes. A slice from a beefsteak tomato warmed by the sun and just picked would cover a whole slab of bread. What fabulous sandwiches these made with mayonnaise. Makes me hungry to think about it.

It was always such fun to visit our garden with Mother, as she would get very enthusiastic about our crops. Most fun was

The very first garden we planted was to grow hemp to serve as winter food for the birds. Estella Jr. is holding the yew bow that Dad had made for her.

to dig potatoes. I might be at the shovel, and Mother with a bucket was collecting the potatoes, which were invariably healthy and robust. We would both get on our knees and feel around for the potatoes. Mother would get excited and ooh and ah about their size, their ruddiness, and their abundance. "Oh, Estella. Look at THAT one! Put your hands through that loose soil and make sure we did not leave any potatoes behind! Those little potatoes are hiding," she would say, or

Bounty from the Shack garden. Mother is collecting some vegetables for a meal.

words to that effect. Boiling these little potatoes up for supper was a special treat, as they were so tasty with butter when lightly cooked.

Our corn occasionally bore enough cobs to give us a meal. We usually had good luck with onions, too.

To weed this garden we had a one-wheeled cultivator with a hoe or blade attached behind the wheel, which was about one foot in diameter. The wooden handles formed a V sprouting from the axis of the wheel. With two hands one would push this wheeled implement between the rows of crops, and it would turn over the weeds, so one could keep the space between the rows pretty weed-free without too much work.

In late summer our prairie plants would still be in bloom: the dark blue New England aster (*A. novae-angliae*), fringed gentian with its deep blue flower (*Gentiana crinata*), stiff goldenrod (*Solidago rigida*) with its golden flower heads, and joe-pye weed (*Eupatorium maculatum*), which has a flat-topped, purple flower. One of my favorites is the boneset (*Eupatorium perfoliatum*). The cool evenings brought fall colors to the sumacs (*Rhus typhina*) along the edge of the pine plantations.

We had wild grapes growing in abundance between widely spaced oaks and small bushes. They were small but with a very interesting, sharp taste. Mother typically made grape jelly with them. In early years we often harvested the wild grapes and took them home to make a cordial. Mother and Dad would put the grapes in a huge flour sack and squeeze the sack to start the juices. Then they'd put everything in a crock that had a tight lid and add a bottle or two of vodka. Dad said that we would keep the crock closed until our boys came back from the war.

That is exactly what we did, and celebrate with the cordial we did when they were safely home again. To be together again was wonderful.

Carl's Hawks

As a young man, my brother Carl became enamored with falconry. It seemed challenging and was historically a sport for huntsmen and leaders. Dad once wrote, "The hawk as a lethal agent is the perfect flower of that still utterly mysterious alchemy—evolution. No living man can or possibly ever will understand the instinct of predation that we share with our raptorial servant."[1]

Dad's friend Bill Feeney was a falconer. He managed to keep a hawk at all times, and sometimes two hawks. Carl asked Bill if he would help him obtain a hawk. Bill agreed, and they made plans to take trips up to the rocky shores of Lake Huron in the fall. The idea was to attract a hawk to a baited trap as it migrated southward out of Canada.

Bill showed Carl how to make a soft leather hood for different kinds of hawks and to prepare soft leather jesses to put on the hawk's legs. This was part of the preparation. In the field Carl watched how Bill hunted with his falcon and watched him get the hawk to "stoop" and kill a pigeon that Bill had released. Carl was very excited about the prospect of training his own hawk.

One fall they drove north to Lake Huron and prepared to camp while they set up the trap on bare rock, where the hawks could see it. The trap was baited with a live pigeon. That first fall they came home with no birds, but the next year they trapped a Cooper's hawk. Carl built a perch in our garage and set about to make his hood. After the hawk got his leather jesses and was attached to the perch, Carl spent a lot of time showing the bird his gloved hand so the hawk would sit on it. He learned the Cooper's hawk call, a long high-pitched and descending whistle.

Carl fed his hawk various kinds of meat, hamburger or chunks of stew meat, while gently whistling the hawk's call. The hawk also occasionally got freshly trapped mice, which Carl called good "roughage."

Carl's Cooper's hawk resting in Madison, equipped with a hood, awaiting dinnertime or an exercise flight.

The next step was to carry the hawk, when it was hungry, out to an open field with a real pigeon or sparrow. He would release the hawk into the air, wait till it circled over his head, then release the pigeon while calling to the hawk. The Cooper's hawk would dive and hit the released bird in midair. Both birds dropped to the ground; the hawk began to eat the pigeon, as it was hungry.

It was quite wonderful to go out with Carl and watch him release the hawk and see the bird climb high overhead with its gorgeous wings and soar above in circles while watching Carl, who held out either raw meat or the promise of a pigeon. Carl would call to the bird (it was quite a fierce, piercing kind of call) and throw the bait into the air or release the pigeon as bait. The hawk would dive to pounce on it, land, and begin eating. Carl would go to the hawk, compliment him, and let him dine on the meat before enticing the bird back onto his gloved hand, where he could hold its jesses.

Carl explained to us that a Cooper's hawk was a member of the accipiter group, and not in the falcon group. When the Cooper's hawk would sit on its perch, it would lift its tail and release a stream of white spoor backward with considerable force. After a few of these, the hawk left ample whitewash marks on the garage wall. Carl hung a sheet behind the Cooper's hawk to catch the spoor before it hit the garage wall. Mother told Carl he had the bird's group name wrong. Rather than an accipiter, the hawk was an "exhibitor."

One day, after a few of these successful stoops, the Cooper's hawk was not hungry enough, and after the release of a pigeon he did not stoop but flew off. No amount of whistling seemed to get his attention. To our dismay, he was gone. As Dad wrote:

"Moreover the hawk, at the slightest error in technique of handling, may either go tame like *Homo Sapiens* or fly away into the blue. All in all falconry is the perfect hobby."[2]

Carl's appetite for owning a falcon was not abated. The next trip to Lake Huron, he and Bill Feeney were able to trap a small falcon, a pigeon hawk, or merlin. He was Carl's favorite. What a beautiful handsome bird: blue feathers on his back, a dark streak across his eye, and a light-colored breast with orange speckles. The training was going along quite well. Again the new bird spent time on a perch in the garage with occasional exercise flights in a pasture.

A neighbor down the street in Madison, Jim Telford, had raised a great horned owl from a baby to near maturity. It was an enormous bird with great ears and dark eyes. Jimmy asked Carl if he would take the owl up to the Shack and release it into the wild. Carl agreed. This meant that the entire family—Dad, Mother, Nina, Carl (with his falcon), I, and the dog—was packed into our old Chevrolet along with Jim's owl. Someone in the front seat held the owl on his or her fist, and someone sat in the backseat holding the pigeon hawk on his or her fist. The dog was on the floor under our feet. It was quite a ride.

On that trip to the Shack, the owl would rotate his head backward to gaze at the pigeon hawk, who was hooded, of course. No matter which way someone turned with the hawk, the owl would swivel his head around 90° to stare at it.

It was a great relief when we got to the Shack. Carl with the owl on his fist went up the hill into the woods and released it. We all ate lunch then, and Carl built a proper perch for his pigeon hawk to mount in the Shack on a rafter or roof beam.

Carl is standing by the tree house holding the great horned owl that he plans to release in the woods. Shack in the background, and the shower can on a pulley appears behind. Photo looks east.

That night the merlin spent the night safely inside. Its new perch was a short piece of four-by-four with a rope wound around the top of it on the beam. The owl, however, came around the Shack, perhaps out of hunger, and perched on the roof in the dark. As the owl got sleepy, he began to slide down the roof, newly shingled with cedar shakes, skidding over the shingles with his feet: *padum, padum, padum, padum, padum*... Then he fell off the roof, and you could hear his wings, *swish, swish, swish... swish*. Pretty soon he was back up on the roof, landing with a *poom!* and the night was quiet again with just the frogs chirping, the spring peepers, down at the river shore. We all went back to sleep. Then it began again, *padum padum padum padum padum padum... swish, swish,*

swish, swish as the owl awakened sufficiently to fly back up on the roof. It would be quieter for a while, then *poom!* We could hear him land again on the roof.

After the third or fourth time this happened, I heard my dad get up and pull on his jacket. He opened the screen door and went out. I climbed down from the upper bunk and followed him out. "What's up, Daddy?" Dad said nothing, but with serious purpose headed down the path toward the river with the flash light.

"What's happening?" I repeated.

"Damn owl," grumbled Dad.

I followed him down to the river. We were both barefoot. When we got to the shore, Dad took out his penknife and cut a couple of willow sticks. Then he rolled up his pajama legs and waded out into the shallows. He proceeded to spear some frogs. Then he would turn and hand one to me. "Here."

"OK, Daddy, what's this for?"

"Damn owl," grumbled Dad.

We turned and used the flashlight to walk back up to the Shack.

Dad walked around the building and came back with the ladder, which he propped up against the roof. The owl was still sitting up on the rooftop.

"Baby," said Dad, "you climb up the ladder and get up on the roof and hand the frog to that owl."

So I did, and the owl studied me carefully. He was very tame, so there was no trouble climbing toward him and offering him the frog. He did not take it, so I laid it by his feet and climbed back down to the ladder and to the ground. Dad and I went back in and each climbed back into bed.

All was quiet... for a time. Then it started up again, *padum, padum, padum, padum, swish, swish, swish... poom!*

The owl was not appeased by the dead frog. Apparently owls do not eat frogs.

After a sleepless night, we all got up and went on with our day. The owl was no longer in sight. We cooked breakfast over the fire in the Shack. Afterward we began to work again on our carpentry projects, fixing the Shack, and putting putty in the knotholes in the wall.

It was a nice day, and Carl fetched the merlin down from the Shack beam perch. He set up a perch for him out on the lawn in the side yard. The merlin was still hooded and tied to the perch.

Suddenly there was a loud screech from the merlin. The owl had appeared from nowhere and attacked. Carl rushed to his side and removed the owl, but the damage had been done. The merlin was fatally wounded by the owl's claws.

Carl fixed a low box with nest-like towels in it and placed the wounded Merlin in it. In the Shack, he hovered over his dear falcon. On Sunday night, the family and the merlin drove back to Madison in the dark, all of us so very sad that this gorgeous beautiful bird was hurt. A day or so later, the merlin died. The whole family grieved. It was a real blow. I think that marked the end of falconry for Carl.

Hunting Traditions

Hunting was a Leopold family tradition, at the Shack and elsewhere. Archery was the most common, but other times we hunted with shotguns, and our family, in various configurations, went on frequent trips.

As part of our early training in hunting with shotguns, we carried a wooden cut-out gun the shape and size of a rifle. We were taught to always keep it pointed at the ground. One never allowed the wooden gun to be pointed at a person. On early hunting trips in New Mexico, Dad and Mother would each carry a 20-gauge shotgun, and young Starker would follow, carrying the wooden gun. I think their early hunting was for quail. Later they hunted for waterfowl, sometimes using decoys.

In Wisconsin, there were frequent family hunting trips to the Riley Game Cooperative, usually for pheasant. Occasionally Dad and the boys would go after snipe, which are fast little birds, hard to hit.

When the entire family went camping, hunting was usually connected to it. The earliest such trip I recall was to Goose Lake, near Madison, for pheasant. We spread out the big canvas tent, set up camp along a little stream, and made a cooking spot, usually with a forked maple stave or two crossed with a maple bough from which we could hang our teakettle. Never mind that it rained that evening and night, and I (at three and a half) had to sleep in the car parked on the other side of the stream.

At dawn the stream had swollen over its banks, and I can recall crying while waiting for Luna or someone to wade across the stream and pick me up and set me down in the tent. It was miserable, cold and raining. The tent flap was open so we could look out over the wetland. At that stage I recall distinctly Dad handing me a cup of cambric tea (hot water with milk and sugar). I was crying. "Here's your tea. Now stop whining!" I stopped.

Later, the sun came out, and things brightened up. I recall playing with Carl in the little gullies while Dad and the others went pheasant hunting.

We siblings each got trained on the very same single-shot 20-gauge shotgun, which was the only gun I ever used hunting. It was a big day for me when in high school I shot my first pheasant. Curt Meine showed me a letter Dad had written to Starker announcing, "Estella had killed her first pheasant!"[3]

Author's first pheasant. On a hunting trip at Riley, the photo records the fact that I shot my first pheasant. Gus is standing by.

Starker, Luna, Carl, and Dad sometimes went out to our bottomland woods not far from the Shack to hunt ducks in the fall. They also went downstream to Plummer's Slough (below Gilbert's farm). This did not involve the dog usually, so it meant that they had to wade out and pick up any bird they were able to shoot. It would make a great supper later on.

Dad did not want to hunt grouse on our Shack land (because these were "our birds"), so he preferred to drive up to Adams County, where he had permission to hunt on someone's woodland. At the Shack area, we took special pains to record and hear our grouse drumming every spring. Their drumming was a nice part of the spring music in our area. It went well with the calls of the owls and the spring peepers.

By then I was in college. Mother and I would go along. Mother would sit knitting at an appointed picnic spot and await Dad's return. At Dad's encouragement, I would carry that single shot 20-gauge shotgun and try to keep up with Dad when Gus flushed a grouse. Dad would be carrying his double-barreled 20-gauge Fox shotgun. Hunting grouse was a thunderous experience. I would be walking through the thick alder brush some distance off to the side from Dad and following the dog (in this case the new pointer named Flick), who was searching for a scent. Both of us were holding our shotguns at the ready. When a sudden roar of wings would mark the instantaneous flushing of a grouse, it created enough adrenaline to scare one into action, but getting the gun up to your shoulder and following the passage of the bird rapidly disappearing from sight in front of you was a difficult task. Dad was very patient with me, and would eventually carry out a shot himself, usually getting the bird. He was really fast and a darn good shot.

When Dad would return at lunchtime, Mother would always greet him with smiles and questions about how it had gone, and he would tell us about the one that got away.

Early Deer Hunting Near the Shack Property

One famous hunting expedition occurred in the area above Portage, Wisconsin. That particular year (1934) was the first bow-and-arrow season to be opened in the United States.

It was after the fall deer-hunting season that Dad exchanged Christmas letters with his friend, a well-known archer, Roy Case of Racine, Wisconsin. Interestingly, the hunt that Dad described was in the same geographic area where the Shack now stands. In a letter dated January 11, 1935, Dad described the events of this hunt as follows:[4]

> Dear Roy,... If we have another open season in Wisconsin, I take it for granted that you will be on deck, and I have been keeping an eye out for a suitable headquarters. The bunch which hunted with me this fall is negotiating for an old barn on the Wisconsin River northeast of Baraboo, and if we can get it we will rig it out as a camp. It will be too small to hold the whole bunch, which precludes me from the invitation, which I would otherwise immediately extend for you to join us. However, somebody may drop out or your own crowd may not all want to go along. We can watch how things go....
>
> ... The party had altogether about five shots, two of which came to me. Nobody had a good shot except myself. My first chance was a running shot at 40 yards on a medium sized buck in rather dense timber. I had to shoot down an opening before the buck. My elevation was right, but I shot either in front of him or behind him.

My second shot was one of which I will never see the equal. The drivers had just entered a piece of bottomland timber surrounded by open fields. Starker and I were on a point of timber projecting into one of these fields at a crossing. We had not yet got set because of an unfavorable wind and because of lack of time. A huge buck and a doe broke cover almost before the drivers entered timber, and came across an open rye field straight at us. The buck was so large that the doe looked like a fawn.

They entered our point of timber on the run and on the opposite corner from where we stood. Starker shot while running, on a slim chance. The buck then circled into the timber behind us and came out on a little ridge silhouetted against the sky-line and stood there. Apparently they always stop and look around before both before crossing an opening and after crossing it. It was a dark rainy day and I estimated the distance to the buck as 75 yards, which is point blank for my outfit. I shot and the arrow went just over his shoulder. Upon later pacing the distance I found it to be 60 yards. This lack of an automatically correct distance estimate cost me this deer, since everything else was done right. I forgot to say that Starker helped to stop this deer by a sharp whistle.

The buck did not jump at the shot, but he did jump when I reached for another arrow. He ran within 10 feet of the tree at the base of the timber where one of us intended to station ourselves. Of course by that time he was in a high lope.

This was the biggest whitetail buck I have ever seen, and he had a magnificent head. I doubt whether there are any such deer in the northern counties because they do not have corn, alfalfa and acorns to grow on.

I had a third shot which I did not take. It was in the same piece of bottomland and hardwood. The drivers reported two bucks which did not leave the timber, so Starker and I stationed ourselves on trails where we thought they would double back. My station was at the convergence of a slough and an open field. Before long a large buck came bounding noiselessly toward me, but instead of entering the field, on the trail, he veered off to the right and before reaching the field, stopped dead still only 40 yards away. A clump of soft maple sprouts stood exactly between me and his forward ribs. I did not venture a shot because I suspected the second buck. I wish now that I had tried to get through the maple. It would have been almost unmissable shot if the arrow had not hit brush.

In general, our hunt confirmed my conclusion from previous hunts; killing a buck with a bow and arrow means going through a series of unexpected mishaps and keeping it up until one of the haps fails to miss.

All of us agree that we have never had a more enjoyable experience. We camped in a tent and were very comfortable in spite of the rain. I can see though, that in really cold weather we might have suffered considerable hardship. Hence the plans for camping in a building next year. Our bunch agreed to constitute a permanent gang for the next year's effort and to spend most of the summer getting ready.

I found my 50 pound osage [bow] lighter than necessary. I am going up 5 pounds next year.

Estella had numerous chances at does close by, but no chances at bucks. The surprise of the hunt was the way she took punishment from weather, exposure, etc. She had a better time than anybody!

The most astonishing thing in your letter is that Ostland, who got his buck, does not intend to repeat. I cannot understand that at all. My conclusion is that there is simply no sport superior to bow-and-arrow deer hunting....

With best regards, A. L.

The members of this hunting party included several friends, Franklin Henika, Mr. Krutzman, Mr. Gunderson, and Ed Ochsner and four Leopolds (Dad, Mother, Starker, and probably Luna). It turned out that that very spring Aldo Leopold purchased the eighty acres of Shack land, and his friend Franklin Henika made plans and bought a piece of property in Anchor's Woods. It appears that they were each serious about establishing deer-hunting camps in the area.

I was delighted to find this correspondence that explained how Dad and Ed Ochsner and all were bow-and-arrow hunting perhaps on this same landscape in 1934. Ed had found the little barn and the Shack property that was for sale in that year. I could just visualize that group of archers doing their best on that landscape. It turned out that there was *not* an open season again the following year, and Dad and our family became increasingly occupied with rebuilding the Shack as a family endeavor.

Dad and Gus

Our first family dog at the Shack was Flicky, a black and white Springer spaniel with curly hair and a bobbed tail. Flicky went with us everywhere: camping in the early days (before the Shack) and later on many weekend visits to the Shack. He was a fine companion, well behaved, and helpful as a hunter and retriever. Flicky appears in many photos during our early hunting trips and family outings. By accident, Flicky got caught in the swinging door between the kitchen and the dining room in our Madison home. I was at school when this happened. It took his life, and we grieved mightily.

Dad and Mother looked about for a replacement dog that could help with hunting. A graduate student, Art Hawkins, suggested that Dad get in touch with Guido Rahr, a Wisconsinite who raised German shorthaired pointers. I learned later that Mr. Rahr was a member at one time of the Wisconsin Conservation Commission. Dad did reach Mr. Rahr, it turned out, and soon found that he had a fine German shorthair named Gus, trained for hunting and in need of a new home. This dog had large, speckled dark brown spots on dappled gray fur and a bobbed tail. We had a new dog.

Gus, resting on the warm rocks at the front door of the Shack. It is a sunny day.

Gus went with us everywhere. Dad was particularly happy to take him hunting, such as at Riley, where he could hunt pheasants. Gus was excellent at pointing and retrieving birds. He and Dad became great friends right away. Dad took pleasure in feeding the dog—often holding the mix of kibbles and milk in his hand, asking the dog to sit and wait, before feeding him. Then he would set down Gus's plate, laughing delightedly when Gus began to wolf down his food. At the Shack in the evening, Gus would put his head on Dad's knee. On cold days Gus liked to sit in front of the fire, enjoying

Dad smiling and waiting to tell Gus it is OK to eat his dinner.

FALL

the warmth. We all got to love Gus, who turned out to be a very fine field dog. Dad remarked several times that "this dog Gus was just a great treat" as a hunter and pointer. He was especially good at pointing quail and pheasants. Gus would get very excited, sometimes trembling on a point.

One time when Dad was hunting with Gus at Riley, Wisconsin, Gus broke his upper leg while running in a marsh. This was a serious blow. Dad took him directly to a veterinarian, who set the leg and gave Gus a metal brace to hold the break. The brace was unwieldy and the leg healed badly. From then on, Gus had a swollen upper leg. Nevertheless, he got around fine. He worked well hunting and retrieving pheasants, grouse, woodcocks, or ducks.

Our Gus liked to curl up in the warm sun in the backyard and sleep. One of our pet crows would eye the dog, gradually hop up behind Gus, and suddenly reach out and pull his tail. Gus would bolt up and look around, bothered but patient. He would see the crow and then quietly settle down on the ground again to sleep. Gus knew by that time that this funny black bird was a family pet. Gus was indeed very tolerant.

Dad would always cook the game birds fast in a really hot Dutch oven with a pile of hot oak coals on the Shack fireplace. He and Mother kept a small shovel by the fireplace for that purpose. Mother would prepare the vegetables and arrange to have the plates warming on the trivet. Supper would be taken on the Shack table, some distance from the fire.

After we washed up the dishes, we would get out the guitar. Mother and I would sing various Mexican songs, after which Mother knitted, and Dad would sit in his favorite canvas chair. Gus's bed was a gunnysack filled with straw to the right-hand

Gus and Pedro the crow in front yard of Shack. Gus was unusually tolerant of this pet bird, who was a great tease.

side of Dad's chair, close enough that he could reach down once in a while and pet Gus.

One Sunday afternoon when Dad's brother Uncle Carl and Aunt Dolores were visiting at the Shack we packed up the two cars and drove back home. But when we got to Madison we suddenly realized Gus wasn't with us. We had left Gus at the Shack! Everyone was mortified, and especially Dad. He just loved that dog. It was already dark and too late to return that evening. The next morning, a Monday, Dad, Mother, and I drove back up to the Shack in the morning, worrying all the way about poor Gus. We could see his eyes on us as we approached the Shack. He was all dusty with ashes, as he had been lying in the outdoor fireplace that night to keep warm, which was smart. We all hugged Gus, and Daddy told him we were so sorry to have forgotten him and that we hope he would forgive us.

One time, when I was about sixteen, Dad decided that we should hunt for deer at the Shack with our shotguns. In previous years, we had not hunted deer with firepower, but Mother and Dad, Starker, Luna, and Carl would take turns hunting deer with bow and arrow. Nina may have participated. This year the boys were not home (two were in the military and one was in Mexico doing research). Dad had purchased heavy-duty shotgun shells. It was a clear, cold day in late fall, and the river was in flood. Dad and I took the duck boat and our handmade wooden paddles and went upstream to what he called Anchor's Island (also now called Gus's Island), about five hundred meters above our swimming hole.

We parked the boat at the south end of the island. Dad said he would stand there near the boat in readiness, while I would take Gus and walk slowly up the river side of the island. When Gus and I reached the upper end, I was to turn and come back down the middle of the island in hopes of driving a deer past Dad.

As I was walking upstream along the river, I heard a shot off to my left in the woods, some distance away. As we learned later, a hunter in our woods had shot at a doe and wounded her. I kept moving upstream when Gus began to wail and holler like I had never heard him do before. A deer crossed right in front of me and lunged toward the river, with Gus jumping at her right side, balefully yelping. The deer was wounded and blood was trickling down her side. It was the blood that set off Gus like that. I could not shoot, as Gus was in the way. The deer escaped by jumping in the river; Gus was right on her heels, baying loudly.

The deer began swimming, Gus right behind her, still yelping. They got to a sandbar downstream where the deer could stand, but the water was too deep for Gus to stand. The deer reared on her hind legs and came down hard on Gus's back. The deer began to swim the river, Gus still following and yelping. I watched in horror as they proceeded across that broad river, swimming some three hundred yards. The deer made it to the other side well before Gus, who was swept downstream, and was across the river below where I was. Gus, being sorely wounded, probably with a broken back, could not get out of the water. He hung there on the bank, still occasionally yelping.

My heart was beating madly now as I turned tearfully and went down to find Dad. The distance was such that he had not witnessed any of this. Dad asked me exactly where Gus was. I pointed to where he was, right by a big log on the bank and unable to get out. Dad took his shotgun, got in the rickety wooden duck boat, shoved off, and began to paddle. I thought, *Oh, my God*. First, that was a terrific distance to paddle alone, and second, the current was sweeping him downstream. I watched, hoping and praying that Dad could make the other side. He finally did (after an hour's paddle), though quite a distance downstream from Gus. He tied up the boat.

I had Dad's binoculars and watched as he slowly walked the long way upstream, carrying his gun until he finally got to Gus. He leaned down to comfort Gus. I heard a shot. Then I saw Dad slowly turn and walk the long distance back down to where he had tied the boat. He got in the boat, paddled slowly across that swift river, and came to where I was standing. "Let's go home now."

Gus, our German shorthaired pointer. This is a sketch I made of Gus on a weekend trip to the Shack.

We paddled wordlessly down to the swimming hole, parked the boat, and walked slowly up to the Shack. I was crying. It was so frightening, so exhausting for Dad, so heartbreaking. Dad took Mother in his arms and said, "We lost Gus," and sat down by the fire holding Mother's hand. There was nothing to say at such a loss. He did not shed a tear.

Dad wrote a touching essay about Gus in *Round River* and how he had been so desperate to help him and instead had to put him out of his misery.[5]

This was one of the saddest times at our Shack. We all felt this loss of such a good and faithful friend very deeply.

In time it was important to replace Gus with another hunting dog. Dad made contact with Guido Rahr and found a new German shorthaired pointer puppy. Reaching into the past, we all chose the name of Flicky. We welcomed Flicky into the family.

Six

The Evolving Archery Endeavors

> One cannot make a gun—at least I can't. But I can make a bow, and some of them will shoot.... When I look at a rough heavy lumpy splintery stave of bois d'arc, and envision the perfect gleaming weapon that will one day emerge from its graceless innards, and when I picture that bow drawn in a perfect arc, ready—in a split second—to cleave the sky with its shining javelin, I must envision also the probability that it may—in a split second—burst into impotent splinters, while I face another laborious month of evenings at the bench.[1]

This passage, from Dad's essay "Man's Leisure Time," seems to sum up why Dad had turned to bow-and-arrow making as a favorite hobby in the late 1920s.

Archery became a family enterprise. Dad loved to hunt, and we all grew up participating in hunting and archery practice at or near the Shack as well as further afield. And it turned out that Mother particularly had an extraordinary talent and skill for tournament archery.

Roving. Dad with bow ready on a roving exercise at Mr. Horsemeyer's farm west of Madison. Son Luna is on the ready behind him.

Artisan and Archery

My father was a skilled carpenter, and probably learned from his father, Carl Leopold, who was skilled with hand tools. Dad's father was the president of the Leopold Desk Company of Burlington, Iowa, and all of Dad's siblings became expert in woodworking.

Dad started making bows and arrows in 1926, when someone gave him a bow stave of yew wood. In Madison he began to shape this stave, and later ones of Osage orange or yew, into beautiful bows. In the basement he kept his giant toolbox of carpenter tools, which he had transported from Albuquerque. We still have that great box and some of his tools—planes, squares, chisels, files, saws, and such. In the basement he also set up a German-style workbench, with a wood vise on the right side, a metal quick-release vise on the left, and a series of peg holes down the middle to hold a block in place for using a plane.

Dad placed his giant staves of Osage orange or yew in the vice and shaped the bow with a sharp drawknife and a wood file. The midsection, where the grip was located, was carefully shaped and made oval to fit the hand securely. To gauge the symmetry of his sculpturing of the bow stave, he hung up a large sheet of brown paper on the wall and placed a hook high at the top, hung up the bow, and, with the bowstring attached to the sculptured bow tips, pulled the string downward so the bow bent, and so he could see if the curve of the drawn bow was evenly symmetrical. With this arrangement he also was able to measure (in pounds) the strength of the bow in full drawn position.

Starker made his own bows, and Mother and Dad each had several made by Dad, each with a different strength. Each of the children had their own arrows (color coded).

When Dad was satisfied with the shape, he glued a square of leather around the bow handle as a grip, carved pieces of cow horn to receive the bowstring, and glued these "knocks" to the tips of the bow. The craftsmanship was immaculate; Dad even mixed his own glue—made of milk and casein—which he called "the strongest glue that nature offers."[2]

His yew target bow had hand-carved knocks, an ivory arrow plate (where the arrow rests), and a leather handle. Mother's target bow, also of yew, had, in addition to horn knocks, an ivory arrow plate, a rawhide backing for greater durability, and a leather handle.

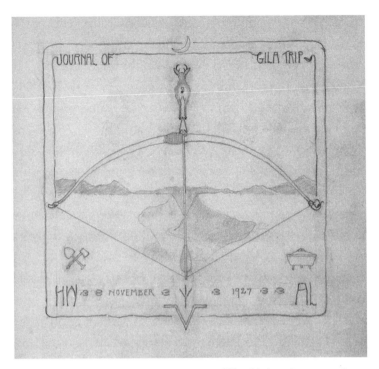

Dad's drawing. He had a brown paper mural like this hanging on our basement wall. There was a hook to hold the bow handle while the bowstring is pulled down to another hook, where a weighing machine records the strength in pounds for the drawn bow.

Dad's hunting bow, on the other hand, was Osage orange wood. He made the bowstrings out of waxed cord, carefully woven together with beeswax. The hunting arrows were made from unfooted Alaska cedar and broad-head points.

Arrow-making involved meticulous craftsmanship as well. When he worked for the Forest Products Laboratory in Madison Dad obtained a motor that he mounted at one end of a narrow bench, and at the other, an arrow-length beyond, he mounted a device that would grip one end of an arrow stave. He ordered rounded staves from a company, but I recall that some he ordered were square staves. I think some of the staves for target practice were of spruce wood, chosen for their lightness. He would spend hours sanding the staves into smooth, rounded arrows using special pairs of blocks with round grooves on them to hold the sandpaper.

As he fashioned the arrows and worked on his bows, Mother was right there in the basement with her knitting, keeping him company. Periodically he would release the clamps, remove an arrow, and, holding one end near his eye, slowly turn the arrow in his fingers to make sure it was ramrod straight. Any with even a slight bend he would throw away.

He was also accumulating in these years a series of classic volumes on archery and arrows, which he housed in a little oak lamp stand at the foot of the stairs. In later stages of arrow production, Dad and Mother would sit in the living room with Mother reading to Dad from one of their favorite books while he worked on gluing the carefully prepared three feathers on the notched tip of each arrow. I remember this period very well. He made an arrow-feather-holding device that could surround the arrow at the notch to hold the feathers exactly 120 degrees apart so they were placed evenly on the arrow.

Roving and Archery Practice

Boxes of arrows later, and with finished bows for which he had calibrated exactly the number of pounds it took to pull the string an arm's length (around fifty or more pounds), he and Mother and the rest of the family (it started before I was born but went on into the early 1930s) would go "roving" (e.g., engaging in target practice and hunting in a pasture). They became acquainted with a nice farmer named Mr. Horsemeyer, who would let them practice bow-and-arrow shooting on his land on the speedway in West Madison. Mr. Horsemeyer's pasture and open woods were attractive for hunting rabbits or gophers. Both Mother and Dad became very good at shooting at targets out there on the farm. Luna and Starker got good at it too. By this time Dad had made bows and arrows for all the children, including Nina, Carl, and eventually even me. I gathered that the boys helped some. They were getting pretty good

Roving. Cousin Bergere Kenney, and Luna and Carl Leopold practicing archery in an oak woodland pasture near Madison.

Starker holding up his trophy, a gopher he shot with his bow and arrow (ca. 1932). Good shooting!

at shooting rabbits and other chosen targets. I can remember eating rabbit for supper, and thought it was OK. There are several pictures in Carl's journal of Starker holding an arrow with a gopher speared by his shot. That is good shooting.

In Madison, Mother and Dad, Starker, and Luna helped to form a local archery club. Dad was proposed as vice president, and they heard Cynthia Wesson, then the women's national champion, deliver a talk on archery.

Mother's Tournament Successes

By 1929 Mother and Dad were having such fun that they attended the Wisconsin Archery Tournament in Racine, Wisconsin. Neither one of them had ever practiced with a full-sized straw target before, yet to their delight and amazement Mother won first prize in the women's division at that tournament—a gold medal.[3] My dad was so proud and pleased. Mother said it was all because of the beautiful equipment that he had built: arrows straight and light, and strong bows with good symmetry. Dad did well but did not place in the men's division.

They next purchased a huge straw target with bullseye and colored rings and a tripod. They got permission from the Vilas Park Zoo authorities in Madison to practice shooting at the park and to store the target in the elephant house when they were not using it. They were soon going down to the park after work with Starker, who was just entering college and had just finished his own yew bow. In the next local tournament in Madison Mom walked off with the first place in the women's division again, and Starker placed fourth in the men's division.[4] Luna and Dad participated as well. Nina shot in the Fort

Atkinson tournament. In 1930, the National Archery Association met in Chicago.[5] Mother and Dad entered, and, sure enough, Mom won three prizes. She won fourth place in the regular national women's contest. She got one first prize for the highest score at sixty yards and another for the "wand shoot" at sixty yards.[6] Now they were practicing many afternoons, and some friends began to shoot with them. The *Chicago Tribune* ran an article about Mother headlined, "Santa Fe Girl Wins National Archery Record."[7] Dad usually placed in these meets, and once he won a silver cup. He was so very proud of Estella!

In July of 1930, the Southern Wisconsin–Northern Illinois State Archery Meet was in Madison. Mother won first prize and also won the "flight shoot": her arrow went 298 yards, nearly equaling the new record of 299 yards (and breaking the previous national record of 273 yards).[8] Apparently she beat the record that Cynthia Wesson had made at that meet.

When I was about three or four years of age I would go with Mother and Dad to Vilas Park in Madison and try to keep out of the way while they practiced bow-and-arrow shooting. It was fun going down to the park after work, with Mother and Dad rolling the round straw targets out of the elephant house and mounting them on their tripod stand.

Sometimes our neighbors Johnnie Emlen, a young ornithology professor at the university, and Virginia, his wife, would join them, having purchased commercial equipment for the purpose.

In the afternoon at the park Mother and Dad would begin by taking turns practicing at a distance of thirty yards from the target. Starker would often join them, and perhaps Luna. Lucky

Johnnie and Virginia Emlen with our family. Left to right, Mother, Starker, Virginia, Dad, and John Emlen at Vilas Park. Flicky joined the party.

boys. People would wonder how my mother was such a great shot with me hanging on to her skirts. I learned to hold my breath and freeze while Mother took aim and fired an arrow. One would hear a *twang*, and the arrow was off coursing its way to the target.

Then, when a quiver of arrows was empty, we would all walk toward the target, picking up any arrows that missed, and pulling arrows from the straw target, with much oohing and aahing at the score. Having warmed up, next Mother and Dad would move back to fifty yards. I am not sure about the hundred-yard practice, but I know that at the tournaments there was a hundred-yard distance shoot, so they had to be familiar with that distance too. It was all part of the activity getting ready for more tournaments... or more deer hunts.

Lady Diana

In 1934 there was a five-day legal hunt in Sauk and Columbia Counties (near Baraboo and Portage).[9] The *Chicago Daily News* pointed out that the sixty-six registered participants were led by "Diana of the Hunt," Mrs. Aldo Leopold.[10] I thought that was wonderful; Mother now had her own nickname in the press. In a personal interview Mother reminded the press, "The stealthiness of the aborigine and all his instinctive lore of the woodland must be mastered by the archer in order to get close enough to deer to bring them down."

The article read, "Not only did Leopold and Estella shoot target together, but they also bow hunted together. Indeed, their relationship attained the ideal. In an unpublished letter Leopold once mentioned that while deer hunting Estella matched the fellows in physical endurance, and generally had a better time than anybody." And the paper went on: "Actually few men have made any contributions to humanity without the assistance of a sympathetic and understanding wife—and in this case a renowned target shooter and bow hunter of the 'Lady Diana' vintage."[11]

By the time Mother had won the state archery championship several years in a row, the Athletic Department of the University of Wisconsin asked her if she would teach their archery class for a year or two. And that is what she did, and she had a lot of fun with it. She and her students would set up their round practice targets along the lawn of Camp Randall near the stadium in Madison. We were very proud that our mother was a part-time faculty member at the University of Wisconsin.

Lady Diana! Mother with bow drawn at archery practice in Madison—probably at Vilas Park. Mother got that nickname in the press when they learned she was not only a target winner but was also a bow-and-arrow hunter!

Mother (*center*) removing her arrows that hit the gold! At the Chicago National Archery Tournament in 1934.

I have always loved the photograph of Mother standing near a target in Chicago with a fistful of her arrows sticking out of the "GOLD" (target) and being interviewed by a reporter at the US National Tournament. She looks slim and beautiful, even having already birthed five children, and she was charming.

Hunting at the Shack and Beyond

From the beginning, archery was part of our Shack experience. In the mid 1936 Dad put up little pegs in a beam in the Shack (to hang the bows on), which suggests that he was planning to do serious deer hunting at the Shack. Everyone did a bit of "roving," but the most excitement was over hunting deer with bow and arrow. It was also a jumping-off point for other archery excursions by various members of the family. For several years, Mother and Dad, usually with Carl, Luna, or Starker, would plan hunting trips at the Shack. They communicated in the field using cow horns that Dad had made; very effective! Hunting with bow and arrow we all found pleasurable, challenging, and often exhilarating. To give you a sense of the excitement on the part of our bow hunters, here's an entry in Luna's journal (below) about a hunt along the Baraboo River, when the first bow-and-arrow season opened in Wisconsin in 1934; there were a total of sixty-three bow-and-arrow hunters registered in the state of Wisconsin at the time. This particular expedition included Dad and Mother as well as Luna and Starker; Carl and Nina were perhaps there too, though not apparently hunting.

> Tuesday [the second day of the outing?] broke with rain and snow falling, and though it was a large bunch to feed, breakfast under the impromptu lean-to went off well. It was with misgivings about our new tackle that we knocked our bows and started out in the wet. Separated in half an hour from the rest (an easy thing to do). Mother and I waited for Daddy in the cleared hollow. I was up on the hill watching for Stark when

I looked down in the hollow and saw three snow-white tails as they disappeared behind the oak brush not twenty yards from Mother. When Stark and I joined her, I could see from a hundred yards the enthusiasm on her face—before we got within talking distance Mother was whispering and motioning, waving her bow. Two does and a fawn had come across the clearing and stopped within 20 yards—didn't see Mother, evidently—wiggled their ears and rainbowed over the ridge.

Daddy didn't show up so the three of us started up the draw edging up the hill. As I walked I saw fresh tracks in the wet snow and it wasn't long before I spied an old man, standing in the dripping trees glaring down the hill at me. In his hand he held a wet looking squirrel. Now Mr. Crane, for I soon learned his name, was a pretty sane individual for a while, but after he had seen our outfit he "let on as how he'd sort of, just for the fun of it, like to see us kill something with them arrows" and he "allowed as how if we didn't go near the house, we could do a little monkeying around", but he was afraid his family would give him the laugh because he swore that "none of them archer fellows goin' to stick his deers full of arrows like a pincushion."

A couple of short drives and not much doing. We saw them stationed in the little glade—*THE* little glade, because it was there that I had a big moment. I had not crouched behind the little oak bush on the sandy open two minutes—still adjusting my position when out of the corner of my eye I saw something going under the fence at the bottom of the draw. No writing can say what went through my mind in the next breathless minutes—alert, trying to watch every place at once—I caught a movement in the brush. A deer was watching me—all I could see was the head, mostly ears. I couldn't see horns—tense, bow half drawn, I watched those ears. They twitched; I twitched.

Move you _____. I let up my bow and never released but my hand was [on the bowstring] when I saw—what the hell—clearing the fence on a dead run and passing thru the open woods. By God—a *buck*! Up at full draw I waited on an opening—and that dam deer *stopped*. Fully covered except his head he looked at me from 75 yards. Well, the ___ he's off—a bound to the next opening—one more—I lead him carefully, elevated for 80 yards and Tssst! A jump, a jump, one more—he's on it!! Right over his back, as pretty as could be, and on he went with a leap—and disappeared in the woods.

This was not the only exciting thing that was going on. Between Mother and Stark, 80 yards from each other. The, and I mean *the*, big buck passed across the open. You can't predict what they'll do.

Coming back to the road in the twilight, two white tails had us on our toes for a few minutes.

Wednesday—Back at the same place and nearly the same drives, when the most exciting drive occurred. Daddy was on the same pass where yesterday my arrow flew, only he stood looking down the lane, and I was posted to the north across the sand blow. The first thing I knew, a little fawn was standing before me in the open at twenty paces, broadside. Maybe I didn't look for horns, too. I thought I heard a bowstring and when I went over to see Daddy, sure enough, a running broadside 40 yard shot at a buck thru an opening—shot 6″ in front of him. Well, it's a great life.

After lunch, Daddy and Mr. Cross [came] from across the road. I had just put the car away, built a blind, and was contemplating the situation when clump-clump-mrrrump. It sounded like a whole remuda of caballos. It seems the wind was blowing down the road toward me—the rest of the outfit was strung downwind from me, along the road. Not being prepared to see deer at my right, I was too surprised to be jumpy and I looked past the edge of my blind at a doe and fawn standing looking back. While I watched, I heard again the clump of hoofs and a big buck swung into the field upwind, crossing at 60 yards. Up I drew—lead—behind a damn oak bush—on him again— 80 yards, more lead, and a year passed as that arrow sped true as I had aimed. They'll meet *by God*—nope, just past his tail, so with an extra jump, the trophy rainbowed across the fence, up over the sand and away. A sight in the world![12]

As Luna wrote, "Deer hunting with the bow is a series of mishaps which eventually will mishap wrong for the buck and he'll be yourn."[13]

Luna's journal entry gives us a taste of the challenge as well as the excitement of bow hunting. One reason the experience draws you in is that it's not easy to bag game that way. Mother had a good eye for the difficulties, which comes through in this expression of her views that appeared in a newspaper account at the time: "We are going for the fun of seeing the deer more than anything else. You see it will be very difficult to get close enough to hit them with an arrow, but we'll try." The account has her estimating that sixty yards was "the maximum distance at which a deer could be slain with an arrow."[14]

The hunting season was a great success. Of the 1934 Wisconsin hunt itself, one press report wrote: "The first and most successful bow-and-arrow deer hunting season in modern times has wound up in a blaze of glory. Prof. Aldo Leopold of the University of Wisconsin announced at Madison today that the 5-day season, just ended, 'was most successful' although no deer were slain. The hunt was so enjoyable that several of the participants decided to make it an annual event."[15]

Occasionally the Shack area in effect became an archery camp. In 1936 a seven-day season opened for bow-and-arrow hunting, beginning October 25. That year several friends of Dad's and Mother's got licenses: Mr. and Mrs. Roy Case and Franklin Henika. Each recipient of a special bow-and-arrow permit had to obtain an upland game hunting license and a deer tag. Only fork-horned bucks could be hunted.

Mother and Dad and Carl had planned a bow hunt for deer at the Shack. This was one of the years in which the Conservation Department first opened the season to bow hunters in Sauk County. Dad and Carl wanted Mother to have a really good chance to get a deer, so they planned this event with care.

It was a nice day. Dad and Carl were spread out in the lowland north of the Terbilcox property east of Lake Chapman and would drive any deer that might be on the lower slopes upward toward where Mother was posted. Mother was stationed on the slope in the oak woodland. Soon she also spotted some lovely wild grapes hanging from a nearby vine. She had just put down her bow and lifted her skirts to pick a whole bunch of grapes when a large buck with a big horn rack walked calmly by at a distance of about forty yards. What a shot she would have had! But Mother, her skirts full of wild grapes, watched with admiration as this handsome beast trotted by her. She could not move, as that would scare the buck. She said excitedly, "Oh, he was such an elegant-looking animal!"

When they met on the hill, Carl and Dad laughed and laughed when she told them what had happened, and how very excited she had been at seeing this big buck glide right by her. They were delighted that "Lady Diana" had had such a fine opportunity—even if she passed it by!

It wasn't just Mother who lost out on some key opportunities, though. On another occasion at the Shack, Dad, holding his long bow, was standing with me on the River Road at the east edge of what Nina and Charley later called the Two Bears Prairie, not far from where an old sawmill once stood. We were looking downstream, east. Carl had been driving game southward from the woods north of the river road. Suddenly a big buck with a fine rack of horns came running southward ahead of Carl and across the great marsh. I immediately dropped to my knees to get out of Dad's way and watched him draw his bow. The deer was about a hundred yards away. Dad took a good aim, and *twang*! He had paced his arrow to be about where the buck was running, but it went about one foot above his back. I was so impressed. It was really an exciting, skillful shot at such a great a distance.

"Oh, really wonderful, Dad!" I exclaimed.

We began walking toward where the arrow landed among the willows beyond and searched for the arrow, without luck. It was a memorable event.

On another occasion, one whose aftermath has special meaning to me, Dad went deer hunting with Starker and Uncle Carl to the Sierra Madre country in northern Mexico. His journal mentioned the music they enjoyed at night in camp, thanks to Floyd, their guide "who knew an infinite number of new Spanish songs." In his journal Dad wrote about "glee club" around the fire at night. On January 11, 1937, he noted, "We are all now working on 'Concha Querida.'" He did not get a deer, but he had a marvelous time and came home from the trip with that song in his head. I was thrilled when he whistled it for me at the Shack, telling me that Starker would have to furnish the words. I immediately got that song going on my guitar while Dad whistled it to Mother and me, and Starker indeed later provided the words. So nice! It is an old song still in play. There are many different songs by this title, but this matches the one Dad gave me. It is documented by Norberto "Beto" Quintanilla (1950s). Here is the first verse:

Concha Querida, luz de mis ojos,
Luz de mi pensamiento
Dale un besito al hombre que te ama
Para que viva contento.

Beloved Concha (shell), light of my eyes
Light of my thoughts
Give a little kiss to the man that loves you
In order to live in contentment.

Seven

The Shack Landscape and Its Restoration: A Natural history

"The outstanding scientific discovery of the twentieth century is not television, or radio, but rather the complexity of the land organism," wrote my father in *Round River*.[1]

As he was hinting, we can locate many of the parts, but how these fit together in the land organism was another matter. Finding the native plant species would be a good start. To reunite some of these came next. The work of our family was creative in its own right: figuring out what conditions these species needed, including by experimentation.

Essential to that is appreciating how this landscape got its form—what processes have worked on it and with what results. This much helps us with our understanding of the setting and the soils—what I would call the lay of the land. In the work to restore old habitats and old vegetation types, it is really useful and interesting to know something of the land history, ancient and recent. As Mary Austin wrote, "To understand the fashion of any life, one must know the land it is lived in and the procession of the year."[2] The Shack experience involved both of these elements.

The Lay of the Land

When you live in an area, a natural question that arises is how the landscape got the way it is. What forces shaped it, and over what periods of time?

In the Shack area, two different prominent ridges (about twenty-five feet in height) are oriented perpendicular to the Wisconsin River. One is the north-south ridge just west of the Shack—the Sand Hill/Clay Hill ridge. The other is the north-south ridge downstream from Gilbert's farm; it is the ridge on which the Leopold Center is built. At the point where the river cuts the nose of that ridge (Barrows Bluff) are a great number of large boulders and clay. The Sand Hill site also has an enormous boulder on it. Both have sand on top near the river.

I wondered how ridges like these formed in the first place. Then I read the report by Robert Dott and John Attig about the history of the glacial ice lobes in Wisconsin.[3] The authors describe how some of the continental glacial ice advanced over eastern Wisconsin and Michigan around 18,000 years before present (BP). Climate changes and floods from ice melt caused various structures to develop at the leading edge of the continental ice. Another geologist suggested that some 18,000 years ago the red clay ridge near the Shack had been clay-rich lake mud that was pushed up by the "Green

Bay glacial ice lobe" when it advanced temporarily from the east.[4] As the climate shifted there were minor pulses of heating and cooling during which the melting ice at the edge of the ice sheet left hummocky ridges of sediment and boulders. Ice advances pushed ridges up, and ice melt during pauses dumped sediment and rock on the land. The Shack area of Sauk County was marked by some of this kind of ridges perpendicular to the river.

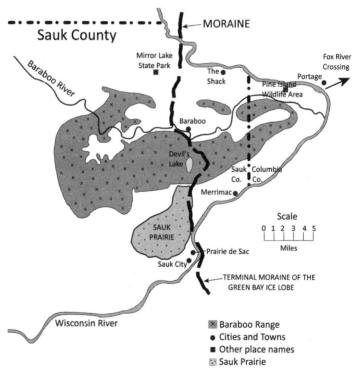

Map 2. Johnstown moraine in the Baraboo Range area. Ice had retreated from this area by circa 11,000 years before present (BP).

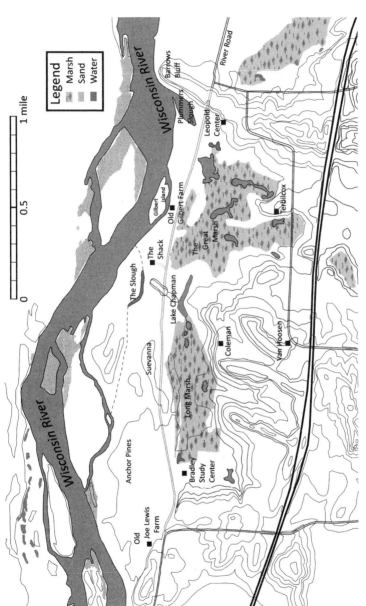

Map 3. Topographic features of the A. Leopold Memorial Reserve (south of the river; after Liegel, 1988).

Glacial Carvings: The Johnstown Moraine and the Green Bay Lobe

One of the major landmarks on the drive from Madison north (on old US 12) to the Shack area is a high, long, north-south ridge called the Johnstown Moraine. It is an impressive pile of sediment that marks the westernmost margin of the Green Bay glacial ice lobe before it retreated from Wisconsin. The ice covered the area between Madison and Baraboo and left this huge deposit of sediments in the form of that enormous ridge, a north-south feature that runs for miles across the state of Wisconsin.

To the west, well beyond the Johnstown Moraine and glacial area is the so-called Driftless Area, the unglaciated part of the Midwest that included most of southwestern Wisconsin and eastern Iowa. Its soils are rich and very old, and its streams are deeply incised in canyons.

The chief sculptor of our Shack landscape, the Green Bay lobe, covered eastern Wisconsin around 18,000 years ago. It is hard to imagine a huge mass of ice, perhaps eighty to one hundred feet thick in parts of eastern Wisconsin.[5] This mass of ice advanced forward several times, responding to climatic changes. Then, as it receded, it left some ridges in our area aligned in a north-south direction, dropping glacial mud and rocks, including granite boulders from the far north (such as Meat Rock on top of the Sand Hill).

It appears that the wide plain on which the Shack and Gilbert's farm sat is the river-washed floodplain between two of those two local ridges. Each ridge is about twenty-five feet in height. The Sand Hill/Clay Hill ridge west of the Shack and Barrows Bluff farther downstream from the Shack. The west-facing slopes of each of these two ridges are

Map 4. Ice of the Wisconsin Glaciation – Green Bay Lobe. (After Dott and Attig, 2004.)

steep, while the east-facing slopes of the ridges are gentle. The gentle slopes may have been where the glacier kept dumping sediment as it melted and retreated to the east.

Both of these north-south ridges are capped with thick sand proximal to the river. I surmise that at some time in the postglacial period, harsh winds swept along the channels of the Wisconsin River, throwing sand up along the tops of the hills near the river, such as at the Shack area on the old sand blow, and probably also on the hill where the Leopold Center is now located (the Barrows Bluff ridge).

Early Forests

What was the original vegetation like? Walk out on a marsh or bog and your feet touch the last peat layers at the end of some old wetland events. You are standing on a virtual history of past records of the marshy basin. Deposits in lake muds and marshes in our area provide an answer, as they contain remnants, in the form of ancient tissues of plants, wood, pollen, and seeds, that can tell us about the vegetation of the time. Geologists and botanists have collected well-preserved pollen, seeds, and wood from sediment cores that provide orderly layer-cake records over time. Plant tissues are notably well preserved in anoxic mud environments. On the marshy shores of Lake Chapman south of the Shack, Marjorie Winkler and Professor Louis Maher of the University of Wisconsin–Madison pulled up lake mud from a core hole. Standing on the wetland at the western margin of Lake Chapman, they drove a piston corer into the marsh, and one meter at a time they slowly pulled up twelve meters of core mud—about three stories' worth. In each segment the proportion of carbon-14 isotopes in the mud told them the age of the sediment (e.g., as old as 12–14,000 years before present). Winkler examined fossilized pollen samples with her microscope at many levels along these cores.[6] The muds from the bottom part of Winkler's core contain remnants of plants that were growing as the glaciers retreated.

For example, the first plants growing on the ice-free landscape were willows, sedges, and a variety of herbaceous plants and small shrubs. Pine and spruce pollen and occasional needles told her that these trees became more numerous in the first thousand years of the postglacial record.

The era of the early conifer forest was also when mastodons inhabited broad areas of Michigan and Wisconsin.[7] Species like black ash (*Fraxinus nigra*) and aspen or cottonwood (*Populus*) were found consistently to be associated with this early conifer forest. That was strange, because ash trees do not now grow with spruce forest. Black ash, a deciduous tree, apparently tolerated the cool temperatures of that era. This kind of forest was widespread, as it also occurred outside the glacial ice area embracing eastern Iowa and southwest Wisconsin. It's wonderful to think about the possibility of mastodons—and perhaps early humans hunting these megafauna—roaming the area near the Shack about twelve thousand years ago.[8]

Vegetation Phases

As the glacial ice melted away from the Milwaukee area, the postglacial warming began in earnest, leading to the temperate climate and the vegetation we are familiar with today. It started with the first vegetation phase, summarized from Attig et al.[9] The very first vegetation consisted of a cold dry spruce-herb parkland followed by periods typified regionally by different forest types:

1. **Spruce Period** and **Pine-Birch Period**. Pollen and needles of the initial spruce/pine forest are recorded in the mud in the cores from Lake Chapman. (~13,000–10,000 years BP). Forests with pine ash and birch (~10,000–9,800 years BP) perhaps represent a drier time. Winkler tells us these were probably white pines and some jack pine as well.
2. **Hardwoods Period**. Between 9,800 and 8,000 years BP a hardwood-conifer forest developed. (Trees can migrate!) The presence of oak, elm, and a number of other hardwoods and pine in the forests midway through this period signify a warming of the region's climate. These species

apparently "migrated" from the southern states where they resided during the worst of the cold Ice Age interval. It took a few thousand years for all the tree taxa to "migrate" (disperse) back to Wisconsin.

3. **Prairie period**. Between 6,500 and 3,500 years BP increased percentages of charcoal (soot) and high amounts of oak and grass pollen suggest that an oak savanna was dominant in south-central Wisconsin. This climate was warmer and drier than now.
4. **Oak Forest** developed in the uplands after 3,500 years BP, suggesting that the climate became cooler and slightly more moist.

Early and Historical Records

The peats at the marsh around Lake Chapman contain fossilized remains of aquatic plants showing the old shorelines and wetlands. During the spruce period the shoreline areas were thick with sedges. In the pine period, water lilies appeared. Clearly the water lilies and sedges that now grow around the lake are really old-timers! (See the water lilies and sedges in photo.) I mean the fossils do tell us they have been growing in the lake for the last ten thousand years. Finally, the pollen spectra in younger sediments are evidence that the Shack area contained a mixture of oaks, hardwoods, and prairie grasses, at a time (ca. 6,000 years ago) when we surmise the native Indians were burning the prairies and woods regularly. The evidence for native burning comes from a series of studies across the Midwest, including the count of carbon particles (soot), which are particularly abundant in the core muds for

Bridge struts at Lake Chapman and Water Lilies in foreground. The pollen and fossil tissue records show that this genus of water lilies has been growing in Lake Chapman since about eight thousand years ago. Camera faces east.

this period about 5,000 C-14 years ago.[10] The pollen evidence tells us that this was the driest period, when expansive prairie was especially well developed in our region and fires were prevalent.

The nature of the association of oak and grass savanna reminds us of the description that John Muir gave of the region of his family's farm in Marquette County, Wisconsin, a few miles north of the Portage area.[11] Muir and his family reported that the original vegetation he and his family found was what they called "oak openings," meaning stands of white oaks or burr oaks (*Quercus alba*, *Q. macrocarpa*) spaced widely apart, with grassy prairie, diverse wildflowers, and small shrubs scattered in between. These oak species had thick bark and were resistant to fire. Standing alone in the prairie, these white oaks were like sentinels. Muir also described how after the Indian burning was stopped the little black and red oaks began to come up on the prairie and took over, making a young forest. The older fireproof white oaks became surrounded. A basic difference between white oak/burr oak group and black and red oak group is the bark thickness of the white/burr oaks in contrast to the thin-barked black and red oaks. The leaves of these groups also differ: white/burr oaks have rounded lobes, while the red/black oaks have sharp pointed lobes on their leaves.

We always dreamed about the historic oak openings and prairie, though we did not know how to rebuild them. When we first were driving from Madison to the Shack in the 1930s, we would drive on the old Highway 12 over the Johnstown Moraine, where there was and is a wide mixture of oaks in the forest. Dad would point out the old, open-grown white oaks surrounded by younger, smaller red oaks and black oaks that came in at the time (ca. 1850) of settlement by peoples of European stock. The structure of these old white/burr oak trees was different. An example of a giant white oak, perhaps 2.5 feet in diameter, exists outside the Bradley Study Center. It shows the features of an open-grown oak, with branches reaching outward. The tree is probably two hundred or more

This veteran white oak is the largest tree on the Leopold Reserve, and it is next to the Bradley Study Center. Lynn Bradley Leopold poses for scale. With a stem diameter of about 2.5 feet, we estimate that the tree is more than two hundred years in age. Its branches are outreaching, indicating it grew in a savanna prairie. Photo courtesy of Alan Anderson.

years of age, so it dates back to when the natives were burning the prairies. In contrast, the branches of the black and red oaks tend to point straight up (seeking light in a forest setting) and tend to be younger.

Why would the native peoples burn the prairies? Well, probably one main reason has to do with hunting and game. The tribes tell us that on the Great Plains it was fire that kept the grasslands open. In the early oak period in Wisconsin the climate was warmer and drier than today, providing conditions for a greater number of natural fires. Indian burning enhanced the grazing for megafauna like the buffaloes, on which the native Indians in the prairie area depended for sustenance and hides.

In one botanical study Jock McAndrews used fossil pollen to reconstruct the nature of the transition (ecotone) area along the prairie/forest margin in western Minnesota.[12] McAndrews demonstrated the nature of the "original" vegetation, and the general effects of burning that kept the prairie open. The prairie and oak-opening vegetation had not changed in composition greatly over the past six thousand years. However, it did change after European settlement in the 1800s, when plows broke the prairie sod, and when oaks of the red/black group began to fill in the meadows between the white and burr oaks and invade the prairie. Invasive weeds began to thrive. The planting of row crops to grow annual plants like rye, corn, and wheat changed the composition of the soil, as these annual plants alone tend to deplete the soil fertility.[13]

In time European agriculture on the savannas began to displace the native prairie species and replace them with crop fields and weeds. A retreat location for many prairie plants became areas with little competition from agriculture, for example, areas along fencerows between fields, roadsides, and along railroad tracks. As mentioned, sparks from steam engines and trains at one time burned vegetation along the edges of the railroad tracks. Such fires reduced the competition for prairie plants, and as a result prairie plants thrived in such areas. We are told that now the railroads are controlling vegetation with chemical retardants, such as Roundup or other products.

Knowing a little about this sequence, we faced the question: How were we going to bring back the native vegetation at the Shack?

What We Did on the Land: Restoration Efforts

In the 1930s, the field of restoration ecology had yet to be invented. Dad was a trailblazer in this area with two projects.

Under his direction, along with the help of Professor John Curtis, the University of Wisconsin–Madison Arboretum was devoted to an attempt to put native species into an old farm landscape. The other pioneering restoration project was, of course, the Shack enterprise and our efforts to bring back the native vegetation of the area. To us it was just fun, and it was exciting to see our transplants coming into bloom and thriving, though of course it was hard work too.

There was no commercial source for these native species. Learning the species was another step. In search of prairie species, we would drive along the back roads looking for prairie plants in bloom. When we saw them along road cuts, we would use a shovel to select examples of these species, put them in tubs in the car trunk, and carry them to the Shack for planting on the old cornfield in front of the Shack yard. In the Portage area was a nice old railroad right of way, and there we found a number of prairie species. We had to do these collections in summer just so we could identify the species when they were in bloom.

Around 1939–40, Dad and Carl were learning the local flora, and Carl was becoming involved in making a "complete" collection of plant species on our Shack land. I think this work was for his bachelor's thesis project in botany under Dr. Norman Fassett. Carl was pressing plants and mounting them on proper herbarium sheets, as well as identifying them by species. That collection is still in the Aldo Leopold Center in the wooden cypress cabinet he made for it. All of us gained from that collection, so it was beneficial for us that he was doing this. We all tried to help Carl get the collections pressed, and learned as we went. Dr. Fassett actually visited the Shack with us once.

During this interval Dad was keeping detailed phenology notes, and the proper Latin names for the species were critical information. He was also working with Elizabeth Jones, a graduate student in botany, to start to develop his phenological publication on bloom dates of the local flora. Mother was learning all these species as well, so it was quite a family topic of conversation.[14]

The Shack Yard—and the Plants We Love

The very first summer at the Shack, we focused our attention on plantings, to decorate the area around it. Some of these plantings were exotic to our area. In spite of our penchant for using only native species, at Mother's behest we first put in a blue lilac bush at the corner of the Shack, where we hung our dishtowels regularly. Mother wanted some *Funkia* plants along the front margin of the Shack. These plants she loved because they decorated the garden of the Starker-Leopold house in Iowa (the plant is still thriving in both places).

So why did we bring in these special plants to decorate the yard? We had a relation—even an emotional connection—to certain plants, and we wanted to have them near us. For one, they remind us of other places, for example *Funkia* in Grandmother's garden in Burlington, Iowa. A book by Patricia Klindienst called *The Earth Knows My Name* beautifully evokes such connections.[15] It describes how immigrant grandmothers brought seeds for herbs, flowers, or special crops from their home countries, intending to plant them in their own garden or yard in the New World. These are familiar plants that we want with us because we love them—in some cases because we need them.

In addition, at the Shack Dad and Mother started a native flower garden. They began planting some especially handsome native prairie species in a little garden right next to the front yard where we could enjoy them. These included such special plants as puccoon (*Lithospermum caroliniense*), the lovely orange bloom Mother was especially fond of. Others included brown-eyed Susans (*Rudbeckia hirta*) and the blue *Liatris spicata* and spiderwort (*Tradescantia* sp.). How we enjoyed the spiderwort. I loved seeing those three-petaled sky-blue flowers, often with dew drops, opening fresh flowers every single morning in summer. Their fresh blue faces greeted you in the warming sun. Of course there were others: blue *Mertensia* was one, a late-spring bloomer. Dad and Mother planted butterfly weed (*Asclepias tuberosa*) from seed in that small garden as well. It provided brilliant orange/red colors in lovely flowers and attracted certain species of butterflies. A cheerful spring flower was birdsfoot violet (*Viola pedata*). Leadplant (*Amorpha canescens*) with its gray foliage was an eye-catcher. All of these colorful plants decorated our near-at-hand landscape at the Shack.

In the little area right in front of the Shack we used a scythe and sometimes a sickle to cut the grass. Only much later did we bring up a hand-pushed lawnmower. The original mower (or one just like it) is still sitting outside the Leopold Center. The scythe was kept under the bunks in the Shack. We sharpened it with a special file. We felt like farmers of old Europe, swinging it back and forth to cut the grass, but it got the job done. It was necessary to minimize mosquito habitat. Mowing around the special garden made it more visible.

The family made a special stop along Highway 12 north of the Johnstown Moraine to find a nice white-barked birch

View of Shack flower garden. Dad and Mother's wildflower garden is between the camera and the Shack. These were special prairie plants that could be admired near the Shack: species like butterfly weed, leadplant, *Mertensia*, and purple coneflower. Robert McCabe and Dad in photo.

(*Betula papyrifera*) to plant near our Shack. We visited a north-facing slope where the white birch was mixed in with the oak woods. From its occurrence one can tell that birch liked the cool north slopes. We picked one out, and this little tree decorated the yard east of the cabin, suffering through warm summers when it probably would have done better in shady woods. I do not think it is there anymore.

Dad and Mother had a special affection for wahoo (*Euonomous afropurpureus*), a beautiful native shrub with striking red fruits in the fall. The species has stems with four little ridges running up and down them so they seem square in section, like wings. The fruit looks a bit like that of the vine bittersweet (*Celastrus scandens*), except its fruits are a flaming chartreuse red (not orange). We found the first plant of this species in the bottomland woods. Wahoo is hard to grow, because deer just love this plant. They like to rub their horns on it and chew it, devastating the plant. So Dad made a cylinder hoop of chicken wire and staked it to surround the bush in our side yard. That served to protect the plant for many years. It was such a beautiful shrub; we wanted it in our yard.

Another favorite shrub was *Amelanchier arborea*, or serviceberry, a member of the rose family. It has pretty white flowers and leaves measuring about two inches in length, and yields pretty berry-like fruits, which are favored by the birds. We now have a big plant five feet tall in the front yard of the Shack.

Out in back of the Shack we put in shade-loving species, like *Trillium grandiflorum* and wood anemone (*Anemone quinquefolia*). Another favorite "small" tree was aspen (*Populus tremuloides*). The petiole of the leaf is flattened, which makes the leaves shimmer and whisper in the wind. The two we planted in the Shack's backyard are huge now and need to be thinned. Aspen is such a beautiful tree with shiny white/gray bark and shoots from the root, so once in a while a group of aspens springs up from the roots of one planting, as they are all connected underground.

We planted a couple of white pines just back of the Shack and admired them as they grew. Eventually they shaded the Shack so much that its exterior boards (on the Shack walls) seemed to change to a darker color, perhaps because they were wet a lot. After many lively family discussions we reluctantly trimmed the biggest white pine, and the Shack's color lightened up again.

A special elm tree grew northwest and slightly behind the Shack. It had a large branch that aimed southward just overhead. One hot summer day, Carl got the idea that we should put a pulley on that branch and hang a shower from it. We used a big oil-tin can about ten inches square and fifteen inches deep, which could hold a lot of water. We punched a lot of holes in the bottom of this can, fitted a two-by-two wood handle

On the elm tree west of the Shack we hung a large can with many holes in the bottom. A rubber mat on a string held the water in until we pulled the string attached to it. A good shower. The treehouse elm with ladder appears in back. Camera points northwest.

across the top, and put a rubber sheet on the inside bottom of the can with a string attached to it. We pumped up pails of water and poured them into the shower can. Then we could lift the can up with the rope through the pulley, stand under the shower, and pull the string. The water would sprinkle down over our naked bodies. It was cold! So there was usually a lot of hollering when the shower was in use. Dad and Mother would stand aside and laugh at this operation, but I don't remember their ever taking a shower like that. We certainly did!

The Old Cornfield, and Finding the Natives

Mr. Alexander's old cornfield stretched out in a southeast direction toward the great marsh and was rampant with weeds (see below), having lain fallow for several years before we bought the property. The land provided an open view of more than a mile toward the distant hills. Dad wanted to keep that view: it was a fine place to watch for wildlife on the land, and the open sky allowed us to see eagles and migrating waterbirds as they came by. Further, as it was unshaded, it gave us a warm place to walk or sit. The cornfield was a perfect place for us to try and build a prairie.

Some of the first plants to show between the old corn stalks were weedy—such as sand burr (*Cenchrus* sp.), which had really tough spines (so you had to be careful never to be barefoot), and other weeds, including peppergrass (*Lepidium*), bluegrass, shepherd's purse (*Capsella*), and sorrel (*Rumex acetosella*). Another was the tiny, inconspicuous annual *Draba*, a member of the mustard family, which my dad wrote about.[16] It is a native, a dear little wisp of a plant that blooms quickly in

Aldo Leopold and students at the Shack yard. Dad maintained the view eastward for good visibility of wildlife. The graduate students are looking across the old cornfield, which is now our new Shack Prairie.

the spring and sets its seed, and then is gone by summer. There was also the native perennial panic grass (*Panicum* sp.), with its tiny nut-like seeds that the doves really like.

There were quail in the old cornfield all the time in the early years. They seemed to like the sandy areas between the panic grasses and probably harvested seeds of the latter. We could hear them talking to each other in summertime, calling *bob white*! We would whistle back, *bob white*! To help the quail,

Starker and Dad early on built a "brush tangle" of elm branches they had thinned from the fencerow near the Parthenon, and planted the orange-berried American vine bittersweet to grow on it. The idea was to make a safe haven for quail, game birds, or rabbits to hide from predators. The vine also provided berries for the birds to eat in winter. Brushy fencerows used to perform this service for game birds and songbirds, but farmers typically have changed over to clean fencerows, using every inch for crops. In his regular *Farm Hour* radio talks on WHA, Dad used to promote the maintenance of brushy fencerows and explain the real benefits for the farmer and for wildlife. In the first two winters Dad and Starker put out feeding boxes with corn or other grain by our "brush tangle," and by this effort hoped to encourage upland game birds on our land.

In our continuing efforts to bring in native vegetation, we transplanted prairie grasses from road cuts (only where it did not seem to matter if we took some), always leaving more of their kind. It was a good way to get some of these native perennials started on our hopeful "prairie." We were especially anxious to get the chief prairie grass, turkeyfoot (*Andropogon gerardii*), but also Indian grass (*Sorghastrum nutans*) and switchgrass (*Panicum virgatum*). In addition there was, among others, white wild indigo (*Baptisia leucantha*), prairie phlox (*Phlox pilosa*), purple coneflower (*Echinacea pallida*), *Rudbeckia hirta*, Kansas gayfeather (*Liatris pycnostachya*), flowering spurge (*Euphorbia corolata*), prairie cinquefoil (*Potentilla arguta*), and prairie clover (*Petalostemon purpurea*). In our hopeful prairie we planted a wide variety of these real prairie plants. They did well and reproduced themselves.

Dad wrote about these as all members of the original prairie community in Sauk County, but gave special attention to the rosinweeds (*Silphium*), which he described in his essay "Prairie Birthday": "Never again," he wrote, "would the hundreds of yellow *Silphium* flowers tickle the bellies of the buffalo."[17] There are two species. One is the compass plant (*Silphium laciniatum*), which has deeply lobed leaves that stand up ramrod straight and always orient themselves north and south—a very reliable compass indicator. The other is the cup plant (*Silphium perfoliatum*), with rounded leaves grasping the stem so they can catch water in their axils when it rains. A good adaptation in a dry environment.

Dad first admired a beautiful tall *Silphium* plant blooming just outside the fence of a small triangular cemetery in Sauk County. Each July he noticed it at the intersection where the road from Sauk City and the road from Prairie Du Sac met on the way north to the Baraboo hills. The plant was six feet tall, with a high sunflower-type flower big as a saucer on top and enormous leaves. He said he thought he would try and dig one up and transplant it. He found one he wanted to move, but its root was huge. After thirty minutes of digging he could see that the thick plant root, shaped like a sweet potato, seemed to go all the way to bedrock. It was then that he decided instead to scatter its seeds on freshly dug soil. That method was a success, as the little leaves indeed appeared, though he had to wait many years for it to bloom. When Nina tried this, her seeds began to bloom after about twelve years. One learns a good deal about these plants when one tries to culture them.

There is a famous paper in *Ecological Monographs* from 1946 by E. J. Dyksterhuis on his search for native plants on the

Fort Worth prairie in Texas.[18] He searched between barren fields and in fencerows for remnant species that had once composed the Texas meadows and fields. He located places where the natives had all but disappeared, collected their seeds, and dug up samples to transplant. In this way Dyksterhuis was able to slowly build up the collection of Texas natives at the Fort Worth prairie. He was doing exactly what we did at the Shack prairie—finding the natives and bringing them back in.

Dyksterhuis was one of the earliest range managers in the Soil Conservation Service. He explained the ecological value of promoting tracts of native prairie perennials instead of weedy meadows that produced little per acre. His work scouting for prairie species scattered on the landscape along fencerows and railroad right of ways was our work too. At the Shack, we brought plants back together that had once evolved as neighbors in a prairie ecosystem. Our goal was to create mixes of native species amongst the pines and hardwoods (oak and walnut) and communities of natives that belonged together to form our own future prairie and savanna.

As our native plants and pine plantation areas grew, Dad began to talk about a possible threat from fire that could wipe out most of the pines. He discussed the importance of building fire lanes that might protect them. Toward that end, we stopped in Baraboo and purchased two long roofing sheets made of corrugated tin. These were about eight feet long and two and a half feet wide. Punching holes in the corners at one end, we slipped wire loops and a rope through to make a lead, so that we could pull the sheets along together on the ground ahead of a fire. Then, on spring weekends when the ground

Starker burning a narrow fire lane to protect the pine plantation from accidental fire.

was dry, we began burning our fire lanes. We chose days with little wind.

Wanting to build a fire lane along the south margin of the birch-row plantation, we walked downwind to its eastern corner and waited for a wind that blew west. We always burned into the wind. We laid the tin sheets parallel to each other about five feet apart. In the morning when the wind was low, we could set fire to the eastern end of the burning strip. We had buckets of water handy, wet brooms, and other equipment to control the burn if it jumped outside of our tin-strip borders. The fire burned slowly upwind, which was perfect for keeping up with the burning zone; it could not go very fast. With members of the family standing by with brooms and equipment, we slowly dragged the tin strips upwind, keeping the burning zone between the tin strips. When we got to the western end we put the fire out. Now we had a nice black burnt zone that could stop incoming fire later in the summer, giving us time to try and protect the pine stand. We thought it would work to stop a fire even if we were not there.

During those first years, we began to notice that the prairie grasses did better on the fire lanes than on the unburned meadow. This was a discovery. It appeared—eventually—that we should burn off the prairie (the former cornfield) every few years. It is interesting that in all the time we wanted to build a real prairie, we did not know how to set fire to it all without endangering our pines, or how much difference it would make if we did. After 1948 when Starker took over directing the management plan for the Shack property, he directed Frank Terbilcox, the caretaker, to do this kind of controlled burning. It was about the time that John Curtis on the UW Arboretum prairie began to actually burn the fields there and wrote about it. Starker got the clue and applied the method with Frank's help at the Shack. Frank's first burn on that prairie generated huge, scary flames, but it worked well. As Frank explained, these occasional ground fires would keep the little volunteer oaks and shrubs from invading our prairie, or at least control them.

We were also anxious to enrich our woods with wildflowers. We brought in wild ginger, anemone, Jacob's ladder, hepatica, and many more. We had supposed that these shade-loving species were a natural original element in our woods, so we brought them in and planted them. In any case they were beautiful, and it was an enrichment.

Frank's first prescribed fire, back-burning on the Shack Prairie in 1971. Photo looks east.

Maples Soft and Hard

Soft maples (*Acer saccharinum*) grew in abundance along the edge of the slough and in our riverine woods. Carl had begun to tap the soft maples near the Shack, collecting sap in the spring. It turned out that the syrup had very low sugar content, but that did not stop Carl from collecting it in pails. He kept a fire going out in front of the Shack for days so that the sap could be evaporated and made into a kind of dilute syrup. We poured this syrup over our sourdough pancakes and told Carl that it was OK but not as wonderful as real hard maple syrup.

Carl and Dad began to talk about planting hard (sugar) maples (*Acer sacharum*) at the Shack. We visited our friends the Coleman family who lived on Maple Bluff, a suburb of Madison along the northeastern shore of Lake Mendota where hard maples proliferated. Some geologists argued that there were hard maples only on Maple Bluff because of an abundance of clay there. But a report from ecologist Peter Dunwiddie indicated that the original prairie fires around Madison came from the west, leaving the northeastern shore of Lake Mendota unburned. In any case, with Tom Coleman's permission, we visited his woods and dug up a number of sapling hard maples for transplant at the Shack.

One hard maple was planted right behind the main yard bench in the Shack's front yard. That tree is now quite large, about twelve to fourteen inches in diameter. The other maples were spaced out above the orchard but close to the old foundation. These maple trees are now mature and can be tapped in the spring, if someone wants to collect maple sap and boil it down. But it takes a lot of vigilance to tap maples, as the buckets have to be emptied regularly every day or so. And then one has to boil down the liquid for days to make a syrup from it.

Our Vegetable Garden and the Original Orchard

On the gentle eastern slope of the Sand Hill there is a wide swath that used to be an orchard and ultimately became our vegetable garden. Its northern limit is marked by a row of mature, or overmature, red cedar trees that had been planted by a farmer, probably Mr. Baxter's grandfather, in the early or mid-1800s. The orchard area contains two ancient apple trees, one of which Mother was sure was a Wolf River apple. The year we bought the place it was laden with delicious apples. We trimmed and pruned these apple trees in the early years. Dad planted a mulberry tree (*Morus nigra*) just outside the old garden on the west side near the Shack. He thought it would provide good food for the birds in the fall.

We asked one of the farmers to plow parts of the open grass orchard that became our garden each spring. It was perfect for planting tomatoes, potatoes, onions, and corn. After Mr. Gilbert's cows one summer had enjoyed eating their way through the garden, Dad and Carl came up with a complicated plan to fence it in. The fence posts were of dry oak, and every other posthole was to be filled with a double oak fence post. We drilled holes through both the oak posts to insert connecting maple poles.

That fence was a *major* effort to build. It took years. You would think we could have just used barbed wire, but no, our fence had to be in an antique style. Indeed it was.

We built an oak fence around our garden, but it took years to complete. The row of elms is in the background.

On a sunny spot on top of the Sand Hill just west of the orchard, Dad had dug out the sparse local plants and sowed prairie seeds in a one-meter-square quadrat. This was a good experiment, as it came up royally with tall turkeyfoot bluestem prairie grass (*Andropogon gerardii*) and *Andropogon scoparius*, the shorter reddish prairie grass, along with several prairie flowers. He showed that sowing by seed really worked. We did not water the seeds or help them in any way. The quadrat just stood out as a square of prairie up there on the hill. A wondrous sight! Prairie grasses growing right on pure sand!

Along the north edge of the river terrace on which the Shack stands we planted a red cedar tree or two near the Parthenon to provide a barrier so the Parthenon would not stand out visually.

While we were planting our pine plantations, we wanted to have a mix, and include certain native hardwoods. So we dug holes within the pine plantations. Into one hole we threw in a hickory nut, in the next hole we threw in a walnut, and so forth. Around the edges of the plantations we planted sumac (*Rhus typhina*) and other colorful bushes that would brighten the landscape in the fall. This amounted to landscaping the old farm, and in only a few years it had begun to look quite diverse and wonderful.

Moist Prairie South of the River Road

A place that was always dear to my heart was the moist prairie on the south side of the river road, just north of the outlet of Lake Chapman. I spent a good deal of time in my youth observing this prairie and sitting in it. It had these tiny nodding wild onions (*Allium cernuum*) only five inches tall that tasted so sweet. There was milkwort (*Polygala*) with its lovely orchid-like flower, tiny sedges (*Carex* sp.), tiny daisies called *Erigeron*, and many others. Mertens cites several more: shooting stars (*Dodecatheon media*), cinquefoil (*Potentilla arguta*), and lousewort (*Pedicularis*) are typical of moist prairie too.[19] After a while we bought this lovely moist meadow from Mr. Baxter, and Dad asked him to continue mowing it for hay every year. Dad thought, correctly, that this would ensure that no trees would grow on the meadow. But in time Mr. Baxter got older and no longer mowed the meadow. Right away the gray dogwood (*Cornus racemosa*) started to come in densely. The problem, as this shrub expanded, was that *nothing* seemed to grow under the gray dogwood thicket, which is extremely hard to remove. A super mower appeared to be the best way to clear it, and one was eventually applied by Steve Swenson, ecologist at the Leopold Foundation. After mowing the shrubs, should the herbs come in again, one could attempt burning. I do mourn the loss of this moist prairie. The answer of course: do not stop the annual mowing. Alternatively: burn the meadow every three years.

The Sand Blow

Originally one could see the bare yellow sand area on the ridge top above the Shack from many points on the farm. The sand blow stood out, as there were no plants growing on at all on an area twenty yards in diameter. I have already mentioned that we planted a few pines along the upper rim of the sand blow. Some of these were jack pine (*Pinus banksiana*), which we soon decided to eliminate. We did want to keep the sand blow open and sunny. However, one weekend Fran Hamerstrom brought Dad two plants from up north. One was *Hudsonia tomentosa*, a tiny sandy soil type otherwise called dune plant, and the other sweet fern (*Comptonia peregrina*). She thought these might be a nice addition to the plantings. Dad planted *Hudsonia tomentosa* in the middle of the sand blow, where it hung on and then began to spread. It is a tiny, sparse plant with baby blue flowers. He put the sweet fern along the south edge of the sand blow, and darned if it too began to spread, a whole lot. The sweet fern is a northern plant, and the Shack area falls outside of its range. In recent years we have worked hard to reverse this plant cover and tried to reestablish the original open sand blow. It has been extremely hard to do this, as topsoil began to form. This was one experiment we wished we had not tried! These two new plants were well adapted to fill in the sand blow. However, we wanted to keep the sand blow open as long as we could for sentimental purposes. We regretted giving them a chance to do that.

There is a lesson in this story. These two species have excellent adaptations for building the soil around them, and by dint of their sturdy, long roots and the shade they provide, they filled in the space fairly easily in about thirty years. We are still paying for that mistake.

Tamaracks

Practically at the very beginning of our Shack experience, we thought we could plant tamarack trees (*Larix laricina*) in the swamp at the base of the Sand Hill. In that area just west of the old row of "elums" is a swampy area full of alder and river birch and soft maples. We felt sure that we could establish a real tamarack swamp there, and that these trees might be able to establish an acid soil for a kind of a bog. But try as we might, the tamaracks did not do much at all. One tree I recall grew to be about twelve feet tall, and several were smaller than that. Some died. Apparently the soil was not acidic enough to start with. Maybe we were supposed to transplant sphagnum moss to that end, but we did not think of that.

"Pines Planted. Do Not Molest"

Over time we planted white pines in the riverine woods on high spots where they would not be flooded. Along the river road margin was one such area. We wanted to enhance the lovely mature white pines at Anchor's Woods that had persisted nearby from years back.

This is a story told by Carl, usually with great gusto: he and Dad were working up on the Sand Hill one summer when they heard the clank of a shovel and people talking. They followed the sound and, coming over the hill, came upon two middle-aged couples dressed in sports clothes. The men wore baseball hats, and the ladies had on brightly colored slacks. They were parked on the river road and were digging up our white pines. "Our blood pressure began to rise," Carl said. Dad addressed them.

"What are you doing?"

A man answered, "We are just digging up these small pine trees."

"Who told you that you could have these pine trees?" replied Dad.

"Oh, we're not going to hurt them," the fellow responded. "We are going to plant them in our front yard, and they will do fine!"

At this Carl thought Dad was going to blow up, but instead he pushed his hat back on his head and said loudly, "Well, god*damn*!"

The man responded, "Please, there are ladies present!"

Where upon Dad began to laugh and laugh and laugh. That was rare!

The next week, Dad asked Carl to paint some wooden signs we would nail to the trees along the river road. "PINES PLANTED. DO NOT MOLEST." As we nailed the signs up near our trees, I asked Carl, "What's that word mean, 'molest'? Will that stop these people, Carl?"[20]

From Dad's point of view our pines had a "biotic right" to grow where we put them. Doggone it, they had a right to stay right where they are!

The Neighbor's Fire, April 21, 1948

One morning in April 1948, Mother and Dad and I were at the Shack with the new dog Flicky, a German shorthaired pointer that replaced our Gus. It was a lovely time of year, with many wildflowers coming into bloom and migratory birds coming through.

Mother and Dad took a leisurely walk that morning toward the area near the old Good Oak stump where Dad had seen a special new wildflower blooming, a *Cardamine* that he wanted to show Mother.

We ate our breakfast and were cleaning up afterward, when Dad noticed a plume of smoke in the sky downriver, near Mr. Regan's farmhouse. Because the wind was blowing our way, Dad began to be concerned. Pretty soon he asked me to climb up into the rafters of the Shack and pull down the fire-fighting equipment—the backpack pump. Thereupon he began to pump water fast into this rig, as though he was really worried. He was pumping hard and told me to get out the rake and some other stuff. I had never seen Dad so frightened. He was really jittery. Mother and I were now really worried, too.

The smoke plume looked even bigger by mid-morning, and Dad said it was time for us to go down and see if we could help Mr. Regan control this fire. We left the dog at the Shack and put the gear in the green Chevy, and Dad drove us down the river road. When we got to a place where we could see the Regan Marsh on our right and part of Gilbert's property on the left, he stopped.

"Now Baby, will you please drive over to Mr. Van Hoosen's and call the fire department for the county, and tell them that under the Conservation Commission jurisdiction, I am calling the fire department to come and help control this fire at Mr. Regan's farm?"

His voice was shaking.

"When you have done that please return and wade out into the great marsh here" (he pointed) "and with a rake stop any flames that seem to approach the river road or that might head up toward Lake Chapman and onto our property."

He told Mother to take a broom and stand on the road at the one place where he thought the fire might be able to cross the river road onto Gilbert's property or ours. Dad then lifted the heavy backpack pump out of the car, and I helped him put it on. Then he was off, walking toward the smoke area with this huge weight on his back.

I dropped Mother off and drove over to Mr. Van Hoosen's. With my heart in my mouth, I called the fire department and asked them to bring help as soon as they could, and told them the location. I then drove back and returned to the marsh and waded out with a rake to where I hoped I could see what was happening on the hill near Regan's house. The brush at the bottom of the hill was burning, but I could not see much more. I stayed out there in the marsh watching closely but hearing nothing.

After about an hour, the smoke seemed diminished. I saw a man I didn't know walking down Regan's hill toward me. I waited for him to arrive, but as he came I felt this terrible pain in my gut and fear in my heart. When he got to me he said, "We want you to return to the house with me, as your Father is not well."

I looked squarely at him and suddenly I knew. "Oh, it is worse than that, isn't it?"

He looked at me, and we began to walk. I just *knew* we had lost Dad!

"Where is my mother?" I asked the man.

"She is back at your cabin," he replied.

Without words we walked to the green Chevy, and I drove back to the Shack. When I got there Mother was lying across Dad's bed with her arm over her eyes. There was no talking to her. It was just silence. A pall. The new Flicky came over and seemed anxious. A heavy weight hung over the two of us and the dog. We felt an emptiness that cannot be described. We could not talk.

After a while, someone, I do not know who, came and drove us and Flicky back to our house in Madison in silence. We stepped out of the car, and passing through the front hall Mother and I saw Dad's secretary, Patty Murrish, and her helper, Barbara Rogers, silently waiting for us. Others were there. Mother and I silently went up the stairs to Mother and Dad's room.

Our dear Mother was in grief and mourning for about ten years after that. All of us were in great grief. The entire family went down to Burlington to help bury Dad, while I stayed at the house with Flicky, taking care of things there for the week. It was just a terrible time for all of us. It was the end of such an exciting kind of life. It now seemed so empty without Dad.

Exactly a month later, on May 21, 1948, Nina was able to come up from Missouri and visit Mother, and the three of us drove up to the Shack. Very brave of Mother. We did not know what to do with ourselves except to continue doing what Dad would have done, exactly what Dad would be doing then. We took notes of all the birdsong and blooms we noticed, for Dad's journal. I found my notes the other day for that weekend:

Friday May 21, 1948
Blue Oxalis, first bloom
White Dodecatheon, full bloom
Crowfoot Violet, full bloom
Quail song
Woodcock Peent 7:45
Red headed woodpecker at nest at Regan's farm.
Nuthatch feeding young?
Bastard's toadflax first bloom

> Partridge drumming many times from 6–7pm
> Columbine full bloom
> 2 black ducks on the slough
> *Saturday 5/22/48*
> Hummingbird in yard
> Lupine first bloom
> Sand Cherry out
> Pileated still at end of elms, nesting?

And it went on from there:

> *Sunday 5/23/48*
> Black cherry bloom
> May Apple bloom

Those plants were going to continue blooming and the birds calling without Dad watching over them. How could *that* be?

We Plant an Oak

A year later, in the summer, Luna was with Mother and me at the Shack. Luna said we should want to plant a real white oak or a burr oak for Dad, if we could find one, and plant it right in front of the Shack—these were the oak types that Dad loved the most. It is actually very hard to tell the two closely related oak species apart in their youth. Both are very special because they were the "fire-proof" oaks that probably stood in the savanna-like original prairie ecosystem in Sauk County.

There was precedent for this planting. In years gone by the famous British ecologist Charles Elton had come to visit Dad, and we took him to the Shack. Dad had kept a supply of acorns on hand that Carl and I had collected in Madison and were planting among the newly planted pine trees. Dad asked Elton to plant one of these acorns in the front yard right in front of the cabin, to commemorate his visit. On another occasion much later, in the summer of about 1952, the famous Scottish ecologist Fraser Darling, an admirer of Dad's and a friend of Starker's, visited the Shack with Mother, Nina, and me. He was such a thoughtful, erudite ecologist and a charming person. Nina and Mother and I asked him to plant one of Dad's acorns, also in front of the Shack.[21]

Both of these oaks came up and are still growing. The ironic part of this was that both acorns turned out to be black oaks, and not the white or burr type of oak that Dad loved so much. (Apparently Carl and I collected the wrong kind of acorns back when in Madison for planting at the Shack). Oh dear. The next summer when Luna and Mother and I moved a lovely young oak sapling from down near Mr. Gilbert's property line, we made sure it was a white-oak type. (Actually we thought at first it was a burr oak, but in the end it *was* in fact a white oak.) That planting was especially important, because it was the *right* species (or white oak group). Dad's connection with the white oaks was strong. It is a Good Oak.

That white oak is sturdy and perfectly lovely even today. I can remember what a job it was to dig up the sapling, already then about six feet tall, with its extensive root system. It took all three of us to lift the sapling onto a canvas and hoist it into the trunk of our Chevy. We drove it to the Shack and dug a huge hole and planted the Aldo Leopold Oak.

Eight

The Continuing Process of Restoration, 1948–Present

The process of restoration of the Shack lands did not end with Dad's passing. Quite the contrary. It was picked up by several of his children and by some key neighbors and Wisconsin-based foundations, and ultimately by the Leopold Foundation staff, which continued by expanding the prairies. In this regard, some special recognition is due my sister Nina and her second husband, Charles Bradley, for their initial work developing new prairie areas in Sauk County. Their methods in building a prairie were novel additions to the work/technology that Aldo Leopold and John Curtis had started at the UW Arboretum in Madison.

The Aldo Leopold Memorial Reserve

During the years 1940–1948, Dad continued to purchase more acres, so that by 1948 our holdings were about 350 acres in Fairfield Township, Sauk County. These acres were all contiguous with the original Shack lands. Nearby, Mother and Dad's friends the Thomas Coleman family had over the years enjoyed the log cabin they had built on their land high above Lake Chapman overlooking the great marsh and floodplain. Reed Coleman, the younger son of Tom Coleman, with conservation in mind, in time wanted to expand the land holdings his father had purchased on the south side of the river road across from Lake Chapman. Reed and his colleague and friend Howard Mead laid a plan for the L. R. Head Foundation to gradually purchase nearby parcels of land as they became available from retiring farmers. The Head Foundation was able to compile a huge protected reserve surrounding the 350 or so acres that Dad had bought. It was a creative effort to protect the land of the region from being degraded by home developers and the like. Over the years from 1950 to the 1970s the Head Foundation succeeded in building what is now called the Aldo Leopold Memorial Reserve. This expansive project served indeed to stave off local development. Reed said that his effort was inspired by witnessing the subdividing of the old Gilbert farm along the river into slices of land for summer homes, and he did not want this to happen around either the Shack or the original Coleman land area.[1]

Reed and the L. R. Head Foundation decided early on to draft a potential agreement among local landowners making them members of a new foundation that would protect the land from development. The signers would "agree to restrict the activity of landowners such as limits on building construction, or wetland draining on the property."[2] Starker took the lead for our family. By signing each would agree to give the first right of refusal to the Head Foundation should any of the owners want to put their land up for sale. Eventually the following farmers signed on to the agreement: Carl and Eleanor Anchor, Russell and Dorothy Van Hoosen, and Frank and Coleen Terbilcox. The Leopold family signed on as the Sand County Trust in 1969. The Head Foundation changed its name to the Sand County Foundation.

Frank Terbilcox, hired by the Head Foundation in 1968, enthusiastically took over management of the land, and was very helpful to Mother and our family. Frank met with Starker and Aldo Leopold's student Bob Ellarson to develop ecological plans for managing the Leopold property, and they also discussed management of the far larger Leopold Reserve. It was about in the 1970s that Frank carried out controlled burns at the Shack, as by this time John Curtis was carrying out prescribed burns at the UW Arboretum. By 1973 the reserve, now called the Aldo Leopold Memorial Reserve (LMR), was being hailed in the press as a "Mecca for Ecologists."[3] It was no small matter, as Reed and the Sand County Foundation ultimately added about 1,500 acres to the Leopold Memorial Reserve. The general area of the reserve is shown on Map 3 in Chapter 7. The history of the reserve has been written up by Stephen Laubach in his book *Living the Land Ethic*.

The Bradley Study Center and a Prairie Experiment

In the 1970s my sister Nina and her second husband, Charlie Bradley, wanted to retire and live near the Leopold shack.

They approached the Head Foundation and requested to "rent" five acres of Sand County Foundation land in order to build a log home right next to the biggest and oldest white oak in the whole area of the reserve, a beautiful old growth tree about four feet in diameter, mentioned earlier. This magnificent old oak tree had outreaching branches that told you this oak grew up in the open, probably not surrounded by any forest. The site may have once been surrounded by prairie, as this big oak was older and bigger than any other tree in the area. This tree may have even shaded the buffalo. It thus had probably once been part of an "oak opening" savanna as described by John Muir at his family farm near Portage.[4] Charlie made an ink drawing of this particular white oak (included below.)

Charles Bradley drew this sketch of the giant white oak in front of the Bradley Study Center (photo in chapter 7).

Charlie and Nina gained the necessary foundation permission and began construction in 1976. In their planning for the local landscape, the Bradleys hired a neighbor contractor to scoop out a deep pond for swimming and boating. In the process of doing this, the farmer spread the sediments from what must have been an old bog swale out over the nearby agricultural field, about a hundred yards by sixty yards. Because of the depth of the sediment excavated below the surface, we knew the mud was ancient and thus did not contain the seeds of modern weeds.

Anticipating this new pile of dirt, Nina and Charlie spent a good amount of time in advance collecting native prairie seeds to spread and reseed this raw soil area. They sowed these seeds on the raw land and hired Frank Terbilcox to drag a log over the soil surface with his tractor to help cover the seeds in the fall of 1976. Mixed in with the prairie seeds were seeds they had added of annual oats (*Avena fatua*) as a kind of cover crop and shade for the new baby prairie plants as they emerged. During the following summer, the prairie plants began to show. Nina and Charlie were all excited to see the new prairie developing. I brought Nina a gadget called a point counter in order for her to measure the percent of bare soil (versus the percent covered by prairie plants) as this prairie was developing. The point counter had ten steel wires extending downward from a horizontal cross bar. One would lower these steel wires one at a time and record whether the tip first encountered the soil, a rock, or a plant, in which case one recorded the species of plant. After three hundred count hits in different places with the steel wires one could determine what percent of the new prairie area was covered with plant or was raw soil. Over the first summer, the small prairie plants began to gradually cover the bare soil.

Nina had included seeds of about thirty prairie species initially; but later she added many other species to enrich the prairie in specified areas marked with a metal rebar stake. It took about five years for the prairie-seeded plants to pretty well cover the raw soil. After seven years these plants had their roots way down deep in the soil and were protecting the land from erosion.[5]

In preparation for building their log house, Charlie and Nina had hired a forester to thin the Leopold pine plantations, and allowed the logs to dry over a year's time. These were not tiny logs but massive trees some ten to twelve inches in diameter. What a wonderful reason to thin the old plantations, and what a fine way to make use of the trees we all had planted as children in the 1940s!

This log home, completed in 1977, now called the Bradley Study Center, on Sand County Foundation land, became a center of activity for the Leopold family. All my siblings and available relatives met there on certain holidays in a happy reunion to spend time outside together and to discuss the future of the Shack land and its restoration. That was the occasion and meeting place of the Leopold Shack Foundation, as it was initially called. Nina and Charlie's first research on the land comprised restoring the prairie vegetation on the spoil pile of the pond they had dug, the present day EBL Prairie behind their home and work center. To this day this prairie is one of the best (meaning most diverse), the most weed-free, and the most beautiful on the LMR and in the region. Nina named it after Mother.

Mother at the EBL Prairie, admiring the yellow coneflowers. Ca. 1970.

An important adjunct to building the Bradley Study Center was the construction of a large garden with a great deer-proof fence just south of the house. The reason of course for such a fence was and is the fact that the deer population in our area is just huge; local gardens are decimated by these deer. Nina immediately planted raspberries, grapes, blueberries, and flower beds in the garden and raised beds for growing a wide variety of vegetables. They lived off of this produce and gave much away to neighbors and friends.

The Leopold Fellows Program

Charlie Bradley was an active geologist who in retirement from a professorship at Montana State University wanted to pursue educational outreach and research on the Leopold and Coleman lands. Nina, a botanist and prairie enthusiast, couldn't wait to get started. First they started the Leopold Fellows Program in 1978 under the auspices of the Sand County Foundation. To support student fellowships, Charlie donated some funds and helped raise other moneys for the Sand County Foundation. The Aldo Leopold Shack Foundation paid part of the manager's salary at the LMR during these years. Charles and Nina became the unpaid directors of research for the Sand County Foundation. Over the years Charlie and Nina directed a total of thirty-three students and other researchers who carried out diverse ecological research on the reserve, ranging from hydrogeologic, paleoecologic, to prairie restoration work, including surveys of mammals, birds, insects, and plants and historical ecology studies of the original vegetation. A series of two PhDs and seven master of science degrees were completed under their direction.

Charlie kept daily detailed and oftentimes amusing journal entries on events and work on the reserve, including seasonal blooming and bird migration records, which Nina continued until 2001.[6] Nina's able assistant Teresa Mayer still continues these same phenology observations for the Aldo Leopold Foundation. Nina and their wonderful student Konrad Liegel, with the assistance of some other Leopold Fellows, had fun establishing about ten reconstituted prairies in Sauk County, which are managed as prairie by occasional burning to this day. Some are along the highway by the Badger Ordnance Works south of Baraboo, and several are not far from the Shack area north of Baraboo. Other Leopold Fellows were helpful too: Charlie Luthin, Rick Knight, Mike Mossman, etc. Konrad Liegel published two constructive historical and ecological maps showing the vegetation types at the Leopold Reserve at different intervals. These are valuable studies.[7]

What with the presence of graduate students working at the reserve, plus ecologists on the local staff of the International Crane Foundation (ICF) and interested neighbors in the region, there was a real audience of interested friends for seminars. So Charlie and Nina developed a successful summer Seminar Series weekly during the summer months every year until 1987; the occasions often ended up with music fests around the campfire at the Shack. The program brought in a variety of neighbors and members of the ICF on a weekly basis. These established constructive contact with the neighbors regionally. From 1987 forward the staff has continued these seminars in the warm season.

A summer seminar in the Shack yard. The staff has continued these at intervals at the Shack. Sometimes the group ended up singing together with Curt Meine playing guitar.

Eventually there was a fair amount of scrapping between Nina and Reed Coleman over the financial particulars regarding the Bradley Center. Inasmuch as Charlie had built the Bradley Center, promised to donate it to Reed, and was paying monthly rent and yearly taxes on the house, I gather that Nina felt that Charlie did not need to continue to pay rent forever. There were some tough exchanges over this matter. Eventually the Aldo Leopold Foundation purchased the Bradley Center from the Sand County Foundation.

The Significance of Prairie Building

There was something very important that Nina and Charlie learned by building that first prairie on the spoils from their new pond. When they sowed prairie seeds there, they began to realize that they had built what I call "instant prairie." Those prairie seeds came up nicely, and in five years they had pure prairie vegetation with *no weeds*! Nina was so excited when the young prairie plants sprouted that she went out there on the new ground regularly measuring the amount of bare soil that was being closed in by these new prairie plants. How wonderful! Their experience contrasted with what Aldo Leopold and John Curtis had experienced at the University of Wisconsin Arboretum. When Dad became director of the arboretum, the emphasis was to bring in prairie plants on sod, with some by planting seed. But it was hard to control the weeds, as the prairie plants had to compete with the weeds that were already there. The arboretum lands had been farmlands, and they supported a hefty crop of weeds. The staff fought those weeds for years, but they were never able to completely eradicate them. However, with burning the prairie every few years ("prescribed burns"), the weed percentage was ultimately kept fairly low.[8]

The weed problem has always been a real one for ecologists in trying to build a new prairie. Nina thought more about the idea of building prairie on land without weeds. As an experiment they built their next prairie, one called Two Bears Prairie, not far from the Bradley Center. The name has a story behind it: there had been an old waist-high pair of black (burned) stumps of soft maple on this grassy plot, and their dog Flicky (another German shorthaired pointer) used to bark furiously at these stumps, as though they were megafauna—maybe a pair of wandering bears! We all laughed at Flick, and Nina christened the prairie "Two Bears" decorated with these black stumps after that idea. In order to control the weeds here, she and Charles hired someone to come in and spray the future prairie area with biodegradable Roundup (glyphosate). At that time this chemical was considered a useful tool for eliminating weeds. However, since then, various real concerns have been raised about the broad-scale use of Roundup, which contains toxins.

After spraying the ground three times in one summer to kill off the weeds, they sowed new prairie seed on land in the fall, land that was now free of weeds. They were able to produce "instant prairie"—meaning in five years they also had prairie that had no weeds. Nina published a short paper on this finding in *Restoration Notes*.[9] Many ecologists welcomed this procedure and were able to use it for small tracts of new prairie. After the prairie plants are firmly established the root competition is typically fierce against any adventive weeds.

The improvement of soil minerals was substantially increased over the years with the introduction of native plants. The Land Institute in Salinas, Kansas, has completed some formidable research about the qualities of prairie soils and their increasing fertility.[10] First they demonstrated the great depth to which real prairie plants push their roots (as much as ten to twelve feet!), which is one reason why the prairie ecosystem is so stable once it is in place. Secondly they showed that perennial prairie plants over time increase the soil fertility by increasing phosphorous content and also carbon content compared with annual grass cropland, whereas annual row crops (e.g., wheat without fertilizers) reduce soil fertility over time. Soil carbon and total nitrogen are significantly greater in perennial grasslands to a depth of 60 cm. It is now well-known that maintenance of midwestern prairie requires regular burning, such as every three to four years, in order to eliminate sprouts of woody species, such as sumac and oaks.[11]

Among their prairie activities, Charles and Nina Bradley began to recognize the difficulty of finding seed for certain rare prairie species, so they established a prairie nursery not far from their log house for hard-to-find prairie plants. The International Crane Foundation was a cooperator, and the endeavor was funded by a grant from the EPA, subsequently supplemented by grants from Charlie's family Friendship Foundation. The nursery grew and was fenced, and it had its own water supply and a treasure chest of about fifty-eight typical prairie species. Seeds were made available to like-minded restoration ecologists in Sauk County and the local area. Though the nursery has become fallow in the last five years, it did serve as an important resource for prairie ecologists in the region.

During the time that Nina and Charlie conducted the Fellows Program, my niece Susan Leopold (then a Leopold Fellow) and Scott Freeman (a fellow at the ICF) decided to do some serious trimming of brush in a sandy little basin we now call the Suevanna Prairie west of the Shack. Enclosed in a small, low, flat sandy area (an old floodplain) were a wide number of young shrubs and oaks coming up among nice patches of native grasses and sandy openings. Susan and Scott decided to clear out all the shrubs and young saplings and open up this area as a future prairie that could be maintained. The two of them did just that, which was no small job. Once that was done the staff set about to burn the grassy area now cleared of brush, a practice they've repeated every three or four years, yielding a beautiful small prairie. The lovely development after that association was that Susan and Scott married, and we all celebrated both their marriage and their prairie.

The Aldo Leopold Foundation

By about 1980 Starker Leopold and us siblings were considering the long-range future of the Shack land. One option discussed was to turn over the Shack lands to the University of Wisconsin for a research site. At about that time the younger members of our extended family, Patricia Stevenson, John Collins, Susan Leopold, and Carrie Nelson, got together and wrote an urgent letter to Starker, saying they opposed giving the Shack lands to the University. They felt strongly that the family should retain the Shack lands and run it as a family foundation. They felt a warm connection to the Shack and its history in our family. These young people had a good idea. Starker responded to their letter positively and set about to build a constitution and bylaws for the new foundation (then called the Aldo Leopold Shack Foundation). Starker recruited the young lawyer Anne Ross for this job. Nina was able to obtain a grant from the Packard Foundation, and the Shack Foundation was able to hire its first director, ecologist Charlie Luthin.

Charlie Bradley's Woods and Prairie

When a field and a nice oak woods just southwest of the Bradley Study Center came up for sale, Charles Bradley purchased it. The woods with many black oaks south of the log house (the Bradley Center) contained a nice mixture of shagbark hickory (*Carya ovata*), walnut (*Juglans nigra*), and a few white oaks (*Quercus alba*). For the specimens that had been cut we counted between 150–200 years of rings on stumps there, so these were fairly old. These were surrounded by a whole lot of younger, smaller black and red oaks (*Quercus velutina*, *Q. rubra*). Based on the new stumps in Charlie's woods, Nina and I figured that originally the east side of Charlie's ridge was inhabited by scattered hardwoods, mainly slow-growing, open-growing hickory, walnut, and white oaks. (As mentioned, "open-grown" meant their branches extended out widely, collecting solar energy; these were typically either white or burr oaks. In contrast, the "forest-grown" oaks, usually black/red oaks, had limbs growing more vertically, like up-reaching arms to find light, affected by the shade from surrounding trees.)

In this stand between the white oaks were the many younger red oaks and black oaks less than a hundred years old, with wide rings and up-reaching branches, which told us that the younger black and red oaks came in much later. An age determined from the stumps can tell us a lot. On the west side of that ridge was a meadow where there were no stumps and no trees; perhaps that area was always open. Perhaps even original prairie?

When we got our next acting director, Buddy Huffaker, in 1996, the Leopold Foundation staff got to work with Nina and others to collect seeds from existing prairie plants for future sowing. Their first big project was to turn the alfalfa field next to Charlie's woods into prairie. They collected seeds in the fall and poured these into paper sacks, which they hung from the ceiling in Nina and Charlie's porch for the winter so the fruits would season and dry properly. They had the field sprayed with Roundup to take out the weeds. The next season, they sowed a rich mixture of prairie seeds on old agricultural fields on other parts of Charlie's land, near the Bradley Center, and along County Trunk T. I have heard that they had fun making this planting near the Bradley Center a ceremonial occasion, with recitation of a poem or two. How nice! Now those abandoned fields are in native prairie grasses and flowers. It is now a true tallgrass prairie, with an enormously wide diversity of flowers, and it is burned every few years to maintain its ecological health. I might add that it is beautiful. The painting by Victor Bakhtin here is entitled *Sauk Prairie Remembered: A Vision for the Future*. Its diversity of flora is spectacular. In the background one can see the

Sauk Prairie Remembered: A Vision for the Future, by Victor Bakhtin. Col. Mary Yeakel commissioned the painting. Used with permission of the International Crane Foundation.

two highlands (Baraboo Hills) separated by the old valley once cut by the Wisconsin River, called "The Narrows." Beyond that point lies the deep Devil's Lake, where once the river poured through this gap. Note that buffalo and elk are a part of the prairie scene.

Restored Vegetation Areas

Over the last four decades a number of prairies, wetlands, savannas, and forests have been managed and restored, initially through work directed by Nina and Charles Bradley between 1978 and 1987. Several of these areas have been burned periodically in cooperation with the Sand County Foundation, working with manager Kevin McAleese. Below is a list of restored sites in Sauk County.

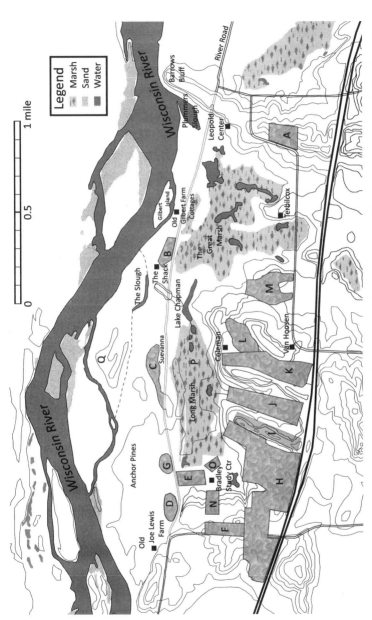

Map 5, showing locations of restored and planted prairies and those only cleared for prairie restoration (all indicated by letters; see text) in the Shack area (map shows the extent of the Aldo Leopold Memorial Reserve).

Each prairie is assigned a letter on the map. Prairies that were cleared and planted by our staff or by our family include:

EBL Prairie at the Bradley Study Center: O
Two Bears Prairie: G
Charlie's Forty Acres, old alfalfa field, 1996: N
Shack Prairie: B
County Trunk T Prairie (on Charlie's Forty): F.
Planted: H, J, K, M
Coleman remnant prairie: L

Other prairies on the LMR are remnants now being managed by clearing and burning:

Suevanna (Established by Scott Freeman and Susan Leopold): O
Frank's Prairie: A
Bradley Study Center: E
Turner Ridge: I
Long Marsh: P
The "Island" area north of the slough which was cleared in 2015: Q

Nearby restored prairies not in the LMR:

Highway 12 Prairie (Konrad Liegel's work with Nina)[12]
Wet Prairie at ICF (Konrad Liegel's work with Jeb Barzen of ICF)[13]
Bent Road Prairie (Nina working with school children)
Corner of Kammerer's wet meadow

The staff burns each of these restored prairies every few years to retard encroaching brush and trees:

Prairie Birthday Cemetery, Sauk County (junction of Highway 12 and Prairie Road from Prairie du Sac)

On the sandstone bluffs surrounding the Sauk Prairie (Badger Ordnance Works area) northwest of Prairie du Sac, Wisconsin, are some steep former prairie areas that recently have become overgrown brush and trees. The blufflands restoration work by staff ecologist Steve Swenson and interns refers to work on the property of several farmers on some of these sunny ridges above Sauk Prairie. The work went on for several years. They have cleared and burned a number of original prairie sites. Because some of these are at the top of steep bluffs and hard to climb, we (Nina, Carl, and I) have called these "goat prairies," a term inherited from our beloved UW botany professor Norman Fassett. This work has brought the restorations goodwill with the farmers and the community. It appeared that the south-facing steep slopes on ridgetops hundreds of feet above the flat land below originally became prairie vegetation because of hot exposures and drought conditions. It was wonderful to see these hot spots cleared of brush and burned so they could return to prairie. The farmers really liked this activity, as it replaced brush with beautiful flowers in the spring.

Oak Forests and Resilient Prairie Plants

An amazing story of restoration and natural resilience revealed itself on Frank Terbilcox's property about a mile southeast of the Shack. One year Frank noticed that his red and black oak woods were besieged with oak wilt. They were on a north-facing slope just above and a bit west of the present location of the Leopold Center building. Frank decided to harvest the dying oaks. In the fall and winter he began to cut up these oaks for firewood, and piled and burned the brush.

To everyone's amazement, the very next summer prairie plants of all sorts popped up on this newly cut area, including tallgrass and medium-grass species and prairie flowers. These plants were blooming and reestablishing prairie in what had been dark, shady oak woods. But *how* did they get there? And how could they pop into bloom so soon after the big oaks were cut? It turns out that these prairie species were lying low (hiding?) in the oak woods, shaded out by these trees, but somehow hanging on by their teeth (as it were) for about sixty to eighty years—amazing! Stories in the magazine *Natural History* tell of the same extraordinary happening in other prairie areas in Pennsylvania, Ohio, and Illinois. The original prairie plants apparently dominated the soil when the black and red oaks invaded the area more than sixty years ago and managed to persist. The oaks grew and prospered. In each case after the oaks were removed, these same old prairie plants were still functioning, and rebounded when they had enough light. What a fascinating discovery!

Of Sandhill Cranes and Ducks

Several wonderful events marked the time when Nina and Charlie lived at the Bradley Study Center. One was the return of the sandhill cranes that Dad had written about, talked about, and described so well. When these once rare birds returned to the Shack area, we celebrated in their honor.

As George Archibald of the International Crane Foundation told us, Dad and others had just about given up on the sandhill cranes coming back to Wisconsin, because their numbers had dwindled so greatly. Dad had spent time with cranes when he visited Clandeboye Bay, the Manitoba marsh on Lake Winnipeg where he immersed himself in the tules and sank neck-deep in the marsh water. There he watched and listened to the wonderful resident grebe, giving her tinkling bell "crick-crock" call. He watched the hen redhead with her convoy of fluffy ducklings ceremoniously following mom, and watched a flotilla of swans gracefully gliding across the waters. Dad wrote one of his most eloquent, heartfelt essays, "Marshland Elegy," as an ode to the cranes, bemoaning how their marshy habitats, like Clandeboye, were being drained to grow wheat. He felt that the beauty and grace of the crane is a quality in nature so special it is "as yet uncaptured by language... as yet beyond the reach of words."[14] I loved his reference to the crane's origin in the remote Eocene, and his notion that when we hear the crane's call we hear no mere bird: "We hear the trumpet in the orchestra of evolution."

In the 1970s on Easter weekend not only did Mother, Susan Flader, and I witness a visit from a pair of cranes coming to Dad's Great Marsh across from the Shack, but, even better, Nina and Charlie discovered in about the year 2000 that a pair of sandhill cranes began nesting in their marsh five hundred yards south of the Bradley Study Center. Now we are fortunate annually to welcome the cranes visiting during migration by the thousands on the shores of the Wisconsin River.

A second marvelous avian event occurred in the first week of May three years in a row. I was visiting Nina, as was Lynn Leopold, Carl's widow. Early one May morning in 2012 I was awakened by Lynn hollering at the top of her lungs to come look at the baby ducks jumping out of the wood duck box across the pond. We rushed to the window overlooking Nina's pond, binoculars in hand. One after another came the baby fluff balls popping out of the wood duck box that stood about fifteen feet above the ground at the edge of the water. The little ducks seemed to bounce on the grass and then scurried toward their mother standing at the edge of the pond. "Oh-h LOOK," we shouted, "here comes another one." This went on until all twelve ducklings were with their mother, whereupon she entered the water, and they all jumped in and swam after her.

This in itself was delightful to watch, but more happened next. The mother, which it turned out was a mother gadwall, was swimming with her flotilla on the pond when an otter appeared from the east, swimming toward the duck family. Mother duck must have said something to the little ones, because they scurried into the shore weeds and disappeared. Mother duck began to fly purposefully toward the otter, paddling her feet along in the water, hydroplaning as she flew. She was talking, as you could see her bill opening. The otter turned and retreated back to the other side away from the ducks, so I guess she scared him enough to make that happen! We cheered and hollered. She then calmly floated on the water, swimming back to her little ones, and climbed out of the pond, the little ones following her, on their way to the next pond.

Amazingly, this happened again in the first week of May the next two years in succession, just at the time of our board meeting at the Leopold Center. What a treat it was!

Nina's Phenology

As soon as Nina and Charlie moved into their new log house, Nina began to set up a system to document phenological observations. Because Nina and Charlie were out and about regularly, they took the occasion to record the first blooms that they saw. These observations particularly focused on the prairie plants, but included all manner of natural history events. After a while Charlie added such events as "date of last swim in October." They recorded when the frogs were singing/calling; the arrival of migratory birds, especially the seed-eaters that came to their bird tray; the freezing-over of the Wisconsin River; and the date when the ice went out each spring. These records got quite voluminous and have been continued right on until today, as Nina's right-hand assistant, Teresa Mayer (a self-made ecologist), who lives nearby, has continued gathering data where Nina left off at the Bradley Study Center.

Nina and Carl had put some of the phenology data together for publication in the *Proceedings of the National Academy of Sciences*.[15] The important finding, as one might expect, was the increasingly early occurrence of events each year, undoubtedly related to climate change. In their paper Nina and Carl showed that, compared with Aldo Leopold's data, the events of 2000 were occurring earlier than in the 1940s. A chart showing the spring bloom and migration dates of arrival[16] presents their data showing that many phenology events are happening about two weeks earlier now than when Dad was taking his notes on this.[17] These events include when the river ice broke up and when Lake Mendota froze over, as well as bloom dates and migration phenomena.

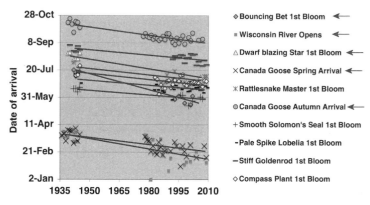

The combined records from Dad's phenology notes and those of Nina show clearly that many blooms and bird migrations were occurring at least two weeks earlier in the 1990s than in the 1940s. Climate warming had begun. Chart used with the permission of David Alan Weinstein, Cornell University.

There is more to add to this phenology story. Teresa Mayer, trained by Nina, is now continuing the phenology observations, as mentioned. Stan Temple, senior fellow at the Leopold Foundation, is writing these up to give us an overall view of the main findings from all these observations at the Leopold Shack area. The staff at the Leopold Center is also taking notes on the blooming time of the prairie plants at that site. So a compilation of this work will soon become available to the public.

The Wisconsin Department of Natural Resources has been publishing a calendar to announce the general blooming dates for many popular plant species during the spring each year.[18] They may be employing Nina's records for the calendar.

Other Restoration Projects

Importantly, the ecologists at the Leopold Foundation, Steve Swenson and Alanna Koshollek, and interns have been continuing a wide range of ecological projects and restoration of the Leopold lands. It is fortunate for us all that Buddy Huffaker, now president of the foundation, has been overseeing a wide variety of projects for the Aldo Leopold Foundation.

One of their most impressive projects is called the Important Bird Area Program. Mike Mossman and Steve Swenson designed a project involving a wide range of lands in Fairfield Township along the Wisconsin River area. The program entails carrying out observations at a wide number (around five hundred) of sites in different local habitats during the spring interval. The observer stands at a station for a certain number of minutes each day during nesting season and records the bird species that are singing or using the area. The observations inform us how diverse the bird community using that particular habitat is. The staff have defined these habitats. When the diversity of bird life is known, it gives the ornithologists a good idea of which habitats are the best for each of these species, and can suggest how best to manage or improve the habitat areas that are less used.

Each bird species requires specific habitat, especially during breeding season. They prefer a certain vegetation type, structure, and local food sources. Also, greater diversity of birds means higher quality habitat (i.e., better land health). In other words, surveying the bird species is a window into the type and quality of the plant communities—land health. It also means our land management objectives can be designed

Our new gate (ca. 1937, *above*). Pine plantings are just started. Later, around 2000, the pines are very tall and spindly (*below*). We learned that we should thin the pine stand vigorously every few years to avoid this loss.

Now we are thinning the pine plantations and some of the woods, upon learning (from the IBA) that the avian fauna that loves savannas will be much happier! More light, more diversity.

for conservation delivery and evaluated through repeated bird monitoring. For example, our prairie restorations of the past were not favored by some expected bird species based on the survey; the foundation now spends considerable time managing the lands between these restorations to create contiguous preferred habitats. Priority bird species for the Important Bird Area were chosen from national and regional conservation plans; successful achievement of our goals enhances these national conservation goals. The lessons learned through this living laboratory and Leopold's own land ethic provide an example and inspiration for landowners far beyond our boundaries.

Our conifer plantings and red/black oak infestations have led to crowded stands, so now our staff is carrying out extensive thinning, opening up stands to improve bird habitat based on what we have learned from the Important Bird Area (IBA) research. We are excited to actively create openings and native savannas and will be using fire as a tool to maintain these good bird habitats.

Driftless Area Landowners

Another key program the staff is engaged in involves establishing contact with landowners in the Driftless Area, a large area of Southern Wisconsin and Minnesota and Iowa that was never glaciated. These are owners who want to have guidance on improving the ecological health of their land. Our staff has been in contact with hundreds of landowners that would like advice on good land management. In fact, our ecologists Steve and Alanna have published several small booklets on ecological guidance for landowners; these are entitled *My Healthy Woods*, and there are individual volumes for various states.[19] These publications focus on building awareness and practical understanding of land care within the capability of landowners. The handbooks focus on values important to landowners such as wildlife, habitat diversity, and land health and are written for the lay person. *My Healthy Woods* has been regionalized for southwest Wisconsin, southeast Minnesota, southern Arkansas, and New Jersey. Collectively the handbooks have reached tens of thousands of landowners owning millions of acres of land. These contacts have led to some great opportunities to help landowners who are seriously interested in land conservation.

Meanwhile our staff is regularly burning prairies, which need attention about every three years. They are conducting prescribed burns with staff and volunteers, and are sometimes

The Leopold Center, a LEED-certified building, built of Leopold logs in 2007, by Boldt Construction of Manitowoc, WI—a great job!

booked by other landowners to come help manage certain private lands ecologically. From time to time, the staff carries out brush removal projects to keep the prairie areas clear of baby oaks and sumac. For two years now, our family has held a kind of reunion on the Shack land and spent an afternoon cutting woody saplings of oak and hazel. It gives our local family and the grandchildren of Nina, Carl, Luna, and Starker a chance to get to the Shack, work together, swim, and then have a party and barbeque. On such occasions we take pride in singing together and hanging out.

Under the capable management of Buddy Huffaker, the Aldo Leopold Foundation board continues to meet three times a year, and embarks on various new projects. Currently the foundation is planning a three-year program raising funds for a Future Leaders Program.

The foundation completed a LEED-certified building, the Leopold Center, in 2006 to serve as our headquarters. It is built on a site along the river road about a mile downstream from the Shack, and is accompanied by a meeting hall and a large garage and work area with prairie plantings. Much of the wood for this construction came from logs of pines and oaks we planted on the Leopold acres.

Nine

The Shack Idea

As each of us siblings—Starker, Luna, Carl, Nina, and I—matured and entered our professional lives in different parts of the country, we carried with us a hankering to have a place in the country, a Shack of our own. It is not merely real estate, of course. Instead, it is a camping place for feeling close to the land, a place to work with the land and to observe the ecosystem and its fauna. To "own," or as the first peoples saw it, to "belong" on a piece of land is exciting and special—a chance to become acquainted with a few favorite species, then to watch them grow. But of course it is way more than that.

As Dad said, he chose his land for its backwardness, but it flourished in splendid isolation under our care. Shack land, as we conceived of it, had the potential of being inhabited by a vast number of native bird species, plus a diverse fauna of mammals, which got richer with time. We were excited that the Shack landscape itself had such physical variety; it had hills and dales, a grand river, a series of tributaries animated by spring and fall floods, a standing bottomland forest coursed by those floods and occupied by lively muskrats, with ducks flying in and out of the sloughs, as well as kingfishers and jays.

Even though it was "degraded" agricultural land, Dad and Mother saw it as a land of opportunities for the family. While it had a "reduced level of complexity," the soil was still there, and we could help improve it, which actually means that the right plants could make it better.

Prairie is the perfect model for this kind of restoration and recovery. Dad described the upward flow of energy from soils through the plant community as a kind of circuit. After major disruption and loss of native species, the energy circuit is slowed and altered. He asked, "Can the land adjust itself to the new order?"[1] He was sure it could if we reintroduced the native plant species on that cornfield, on that terrace, on that hill, in order for a genuine prairie, with its very efficient energy-flow, to become reestablished. To a degree we implemented that recovery; for example, just locating the diverse native prairie species was a needed step. Another was bringing them together. It was Starker's leadership that led us to the necessary periodic burning. Dad's quadrat experiments—such as raising some species from seed and moving blocks of other species to the Shack—provided a progressive step forward. Nina's "instant prairies" yielded the amazingly complex prairie vegetation that now characterizes her back forty. After the first seven years the phosphorus content of the soil at EBL Prairie, as tested by Charlie, had increased several-fold, and this signified a greatly improved fertility.[2] That improvement is thanks to the long network of prairie roots pulling up good minerals.[3]

The delight in experimenting with these species—the pines, the grassland taxa, the forest species—had a tonic effect on us. We saw what was there, we imagined what it could be (or at least Dad did), and we worked to bring it into being, a labor of love. I do think that it was partly this work effort that brought us emotionally close to the Shack land.

The Results: A Mosaic

The pines we planted are beautiful and huge now. There is a bit of natural reproduction by volunteer seedlings in the stand. Several loads for log cabins have been secured from our childhood plantings, and we need to thin the stands further. The prairie on the old cornfield, which is about eighty years old now, and especially the beautiful EBL Prairie (Mother's prairie), demonstrate clearly that we have in fact restored a resource. Those prairie roots probably go down ten or more feet by now! The sod is so tough one can barely get a shovel into it—a big change from our inaugural visit to the place. These are of course features of grasslands that make them stable during drought conditions. The tall grasses and the varied flowers are a beauty to behold. The bird studies the staff members are carrying out inform us that our avian diversity is increased from what we knew in the 1930s. There are species we have lost, of course, and I regret the loss of the bobwhites. Tallgrass prairie is just not their habitat. But with new clearing and innovative management, perhaps the quail will return.

Never mind the delight in witnessing and welcoming all the spring migrants, the experiences banding the chickadees and feeding them at our lunch spots in the woods. Never mind the

Tallgrass prairie once "tickled the bellies of the buffalo." Photo by Nina Leopold Bradley at the Bradley Study Center.

fun we had tracking each other in the winter woods. Or the feeling of accomplishment when everyone helped rebuild the Shack. That made us proud. The idea of making do with the materials around us—the delight in using our hand tools to rebuild the Shack and make it nicer—in fact learning *how* to do all this kind of thing, taught us a lot. Yes, the experience was a creative one, and we learned much about ourselves.

There was something about getting to know the plants and fauna around us that was similar to kinship. In a Shack setting, welcoming the locals as they appeared or reappeared in the different seasons was joyous. It had an emotional twist.

In leaving home we siblings remembered our love for the Shack land; we each sought to find a way to continue our respect for and physical contact with the land and its biota. We wanted to have access to land where we could explore and be close to nature, use hand tools, exercise, and periodically retreat from "too much modernity," as Dad once put it. Having a shanty and land access gives one a chance to follow nature and the seasons in the exact same place during the year—something that cannot be accomplished by a series of field trips to multiple places. In this manner we could witness the ways the ecosystem responded to the seasons. Acquaintance with the same animals, birds, and plants over the course of time also gives one a pleasant feeling of belonging, and of being a part of a natural landscape.

Starker's Place at Sage Hen Field Station, California

The University of California's Sage Hen Field Station, which Starker helped to establish in 1951, became a summer research and camping site for Starker. The field station is in the Sierra Nevada foothills on Sage Hen Creek, near Truckee, California. The elevation is 6,375 feet. My brother was then a professor at the Museum of Vertebrate Zoology at the University of California, Berkeley. Starker built himself a camping site, which became his outdoor cooking operation and a place to keep his camping equipment on a slope overlooking the valley at Sage Hen Creek. His "Shack," as it were, was closely involved with his research in fire ecology and ornithology. At first it was a campsite consisting of a wall with diverse shelves on which he kept his Dutch oven, kettles and pots, utensils, and staples like flour, sugar, coffee, salt, and pepper, along with his bedroll and other equipment. He designed the wall with hinges so he could fold it down for the winter. In the summer, his larder was protected

Starker Leopold in the field (1960s or early 1970s).

there. Out in front he constructed a table and benches where he could get set up for cooking on the open fire, dining, writing in his journal.

In 1965, when Starker became director of Sage Hen, he had a simple cabin built behind his campsite, a place in which he could keep his sleeping equipment and gear. According to his former graduate students Carl and Jane Bock, the camp was in a forest setting with easy access to nearby research areas.[4] Starker's camp was a fine place to talk over ideas and lay out plans with students and colleagues. The climate at Sage Hen is summer-dry, so it was a pleasant area for camping and doing field studies.

During the field season in summer Starker could settle into his campsite, along with wife, Betty, and their son, Fritz, and sometimes his daughter, Sally. They would occasionally invite students and faculty to join them in a meal of sourdough pancakes or steak, and it was a place from which he could operate between fieldwork expeditions. Starker had his guitar up there too, so he could have fun playing his favorite music.

The Sage Hen campsite and cabin served Starker well over the years. Fritz Leopold and his mother would often go to Sage Hen with Starker when he had only a light teaching load. Fritz helped Starker's students with their research—and with live-trapping chipmunks, mapping their movements, and plant collecting.

The topics Starker and his grad students were studying included the relation of deer foraging and grazing to vegetation type and fire history of the land. Their work demonstrated that periodic fire in a summer-dry climate like California's stimulates biotic diversity. The work they completed there ultimately demonstrated the ecological importance of fire as a management tool, and it guided Starker's later assignment by Secretary of the Interior Stewart Udall to help formulate and establish management policies for the National Parks.

In one study his former student Richard Taber was able to discern the distinct preference of mule deer for forage grazing on areas that had been recently burned (within the previous three years). Taber's measurements demonstrated that levels of nitrogen and minerals were notably higher in recently burned areas.[5] Perhaps the forage tasted better as a result. Another student, Carl Bock, carried out detailed studies of woodpecker use of ponderosa pine habitat, especially in relation to fire. Botanist Jane Bock was studying grassland and a fire-dependent system at Sage Hen. The Bocks and Starker's other students went on productively in their professional lives.[6]

Luna's Place on the New Fork, Wyoming

My brother Luna Leopold's refuge was in a different setting. Luna, the river man, had been studying the capacity and characteristics of a small stream tributary, the New Fork, located in the Green River basin of western Wyoming. Having set up a gauging station early on and a contraption to collect and measure sediment carried by the stream, he looked about for a piece of floodplain land near Pinedale, Wyoming, on which to build a log cabin. He wanted ready access to his field sites, but also a getaway place near the mountains for himself and his family. Earlier, Luna had rented land from the US Forest Service on a slope overlooking Fremont Lake, at about 8,500 feet in elevation. This mountain site was well above Pinedale and a place where he and his students could build cabins. This was a vigorous exercise. He purchased two cabin "kits," and one at a time he and the students constructed cabins. Of course Luna designed and built fireplaces in each. They constructed a third cabin using logs obtained from an early settler's cabin, dismembered and relocated to the higher elevation site. A nearby spring served as a source of water. These were primitive cabins.

Luna Leopold in the field in Wyoming. He is sitting on a high terrace of the Green River. Photo courtesy of the University of California Berkeley.

Luna's New Fork log "home," or field station, in the lowlands was to sit on a twelve-acre site he purchased from a rancher. He and his students poured a foundation, and here again he utilized logs from old cabins in the construction. The New Fork log house has two bedrooms, a real bathroom with an old fashioned tub with four "feet," and electricity throughout. A fireplace faces into both the master bedroom and the living room (a special feat of engineering). There is an upper and lower bunk slipped in near the bathroom for visitors, and the kitchen faces east, onto the New Fork stream. It has a narrow back "porch" consisting of huge beams and benches for admiring the stream and a full view of Fremont Peak and the Wind River Mountains. Flaming red wallpaper covers the walls in two rooms; the others feature raw log-faced walls and a good deal of really first-class paintings purchased in Jackson Hole and other art centers.

Out in back Luna constructed a kind of garage and workshop in which he kept saddles, a tool bench, and field equipment.

Luna's Wyoming log house was a great place for him to do fieldwork and fly-fish. It became a center of activity with colleagues, foreign visitors, and friends. Luna and his second wife, Barbara Nelson, and their children (and many of us relatives), along with some of Luna's grad students, helped with the construction, shingling the roof and electrifying the place. Colleagues remember their visits there with enthusiasm and pleasure.

Luna published a series of papers on his work on the hydrology of the New Fork tributary. His most eminent work dealt with the physics of meandering streams and rivers, some of it based on the New Fork. His signal work on the ethics of stream management is a colorful essay published late in his life.[7]

When I visited there we had great sport swimming in the New Fork (very cold!) collected and identified butterflies (Luna's hobby), and talked shop. We also sang various songs from our Shack days and played the guitar around the fireplace at night. There were chances to float the local streams and enjoy the local bird life. Making it especially nice was the appearance of sandhill cranes that used to feed in the pasture next door, along with their baby cranes, who were learning how to make a living. I can remember how Luna got all teary once, watching one poor young crane who had been banded and tagged by the Fish and Wildlife Service with a colored plastic scarf around his neck. None of his fellow cranes, including his siblings, liked him. My big brother had a soft heart for these birds.

Nina and Charlie's Place near the Wisconsin Shack

I've already written about Nina and Charlie's work at the log cabin that became the Bradley Study Center (see chapter 8). Her "Shack story" involved using restoration methods on the local vegetation.[8]

Nina and Charlie worked hard to build an eco-friendly house with appropriate solar panels and shingle roof. This house had a large study/work room in the basement with a wood stove. The house front was faced with gorgeous local rocks. Between the garage and the main house was a mudroom

Nina and Charlie at their woodpile, Bradley Study Center (1980s).

The Bradley Study Center, built of logs from trees we planted as children. A winter view.

with a carpenter bench. There was a screened-in porch and seed room, and a small greenhouse cove. It had a composting toilet that worked very well. Though this log house was no "Shack," it gave them access and proximity to the Leopold Shack, where they had lived during construction and which they wanted for special weekends. The site is a half mile from the nearest neighbor and a mile from the Shack.

As I've mentioned, their published notes on prairie restoration stand out as innovative contributions. There are about 250 known species of prairie plants in the southern Wisconsin area, and many of these—perhaps 150 species—now exist in Nina's special EBL Prairie.

One summer friends and I drove across the Great Plains, taking back roads in order to look for prairie plants. It was the time when the tall compass plant and other rosinweeds were in full bloom in Wisconsin. It should have been a time when one could see such tall plants from the car as we drove through the Great Plains. To my disappointment I could see no prairie plants at all along these roads. Then we arrived in the Black Hills and walked in the fields of the Badlands National Park looking for prairie species. Sure enough, we did find a few scattered prairie grasses (e.g., *Andropogon scoparius*) and brown-eyed Susans (*Rudbeckia hirta*), but no really impressive stands of prairie plants such as Nina had growing on her restored prairies. I came to the conclusion that real prairies were scarce as hen's teeth in the tallgrass prairie region—a sad discovery. British ecologist Fraser Darling, upon taking an ecological tour of the United States, came to the same conclusion. Darling said, "A century ago there were millions of acres of this wonderful vegetational complex, so delicately balanced in relation to buffalo and Indian. Now, even the groping ecologist cannot find a considerable patch of pristine prairie for study purposes."[9]

The Bradley Center became an important gathering place for our extended family. There were many Christmas holidays and Thanksgiving dinners shared there, especially using the rich garden produce that Nina and Charlie harvested over the year. Nights there and at our Leopold Shack were typically times to gather in front of the fireplace, bring out the guitars and sing our old songs, drink wine, and share stories. The young nieces and nephews began to realize that they too loved having access to the Leopold Shack and a chance to revisit old haunts there, welcome back the migrating birds, visit the peenting grounds of the woodcocks, and enjoy the pine plantations, now very large and healthy. With Nina's passing, the

Bradley Center, which is owned by the Leopold Foundation, is now the home of our president, W. Buddy Huffaker, who lives there with this family. His children are happy residents, lapping up the natural history of the place.

Carl and Lynn's Shack in Costa Rica

My brother Carl's shack area was a completely different model. When Carl and his second wife, Lynn Bradley Leopold, visited Costa Rica on a birding vacation, they learned there was a big problem involving tropical rain forests that had been cut. After visiting the mature primary forest areas where the birding was so exciting, they noticed, sadly, that the big open "pasture" areas where the rain forest had been cleared were barren fields. Once used for grazing cattle, these pastures were no longer productive, as the nutrients in the tropical soils had been leached out by the abundant rainfall and could no longer even support cattle. Carl and Lynn joined forces with several other individuals from the Ithaca, New York, area (Carl was a professor and scientist at the Boyce Thompson Institute for Plant Research at Cornell) and established a not-for-profit organization called the Tropical Forestry Initiative (TFI), in an effort to restore former wet tropical forest. They pooled their resources to purchase a *finca* (farm) of about 150 acres and later added another 150 acres. Their purpose was to find a method of replanting the tropical rain forest. Of course everyone said it could not be done! The head of the herbarium at the University of Wisconsin, Dr. Hugh Iltis, said Carl was wasting his time. It was not doable! Well, surprise.

A. Carl Leopold, well-known plant physiologist and ecologist. His work on tropical rain forests and seed physiology is outstanding. Carl's seedling rain forest trees are growing five feet every year now in Costa Rica. This tree is one year old.

When the tropical trees bore their fruit during the winter months (December through March), Carl and Lynn got on their hands and knees and collected the fresh fruits and seeds in the "primary forest" that still existed in the draws and on other properties. They collected fruits from thirteen species of tropical legumes, which, when mature, would become choice marketable lumber. Their hope was that the local inhabitants could selectively harvest them and make money to support themselves. The TFI group named the property Los Arboles ("the trees").

Carl and Lynn began setting up camp in a primitive shack on the property by stretching a large shade cloth between some trees with ropes to form a temporary *plantero* (nursery), shading the tiny seedlings. They set up a gravity-fed water source from a stream. They purchased little plastic growing tubes like the timber company Weyerhaeuser uses to grow Douglas fir seedlings from seed, and put soil and a tropical tree seed in each, placing these in the *plantero* and providing a water source. These seeds began to sprout, and by February they had little leafy tree saplings about a foot tall, ready to plant on the open pasture.

At this point they hired their neighbor, a man by the name of Memo Fallas, to bring his sons and their burro and help plant these little rain forest trees. They used an existing trail for the first set of trees so they could walk it and number individual trees along the transect for monitoring. Elsewhere on the property the tree seedlings were planted well apart in mixed-species plots. The access trail gave them a place to relocate individual saplings to measure their growth each year.

The first three years were remarkable. I was able to visit them in Costa Rica and help measure the trees along that transect and trail. At that time the saplings were between three and five feet tall—and just amazing. The planting operation went on each year for several years. While the first plantings were of species that liked open sun, the later rounds utilized shade-tolerant species that could be planted in the shade of the growing young forest. To our wonder, these too began to flourish and grow tall.

To make a long story short, some of Carl's colleagues from Cornell and Binghamton Universities came down to Costa Rica with a few graduate students to study the new rain forest stands. A facility was built. Now, almost twenty years later, the forest is tall (each tree growing about five feet per year!), and the bird fauna has become more diverse in the new habitat. Toucans, parrots, trogons, and many other forest species are using this new rain forest. Monkeys, sloths, peccaries, and some cats such as ocelots have spread from the primary forest in the deep valleys to enjoy, occupy, and feed in the new forested areas.

The local farmers became interested in the experiment and began to buy seedlings that Carl, Lynn, and the other partners had grown in order to plant them on their own property (as a future investment). This, Carl said, was exactly what they had hoped would happen.

For a number of years Carl and Lynn and the TFI produced and ran a newsletter (called *Tropical Forestry Initiative*) to report on the progress made on the Los Arboles project. They mailed the newsletter out to their friends and colleagues. One of the interesting things about these projects was that Carl brushed up his Spanish and delivered seminars and talks in Europe (but also in the United States) on the wonderful

A. Carl Leopold

discoveries they were making. Another feature was that when I visited him and their shack at Los Arboles, we played the guitar and sang our old songs that we shared at the Leopold Shack. That made it feel like home. Carl's children and grandchildren have had a chance to visit this extraordinary setting and to rejoice in the return of diverse nature, rain forest, on the old barren grazing fields of Los Arboles.

In later years the TFI faculty from Ithaca developed outbuildings for visitors and brought more students down to study the new rain forest ecosystem as it grew. After a number of graduate students took training there, the project closed down for lack of money, but the achievements have been spectacular. Carl published a couple of reports on this project: an important one is titled "Attempting Restoration of Wet Tropic Forests in Costa Rica," and another "Toward Restoration of a Wet Tropical Forest in Costa Rica: A Ten-Year Report."[10] We are very proud of this accomplishment.

My Shack West in Colorado

My own Shack experience differed from the others. As a botanist I took a job as a paleontologist with the US Geological Survey in Denver in 1956. I immediately began to scout for land in the Rocky Mountains to establish my own shack. It took a few years, but eventually I bought part of an old ranch about fifty miles outside of Denver (Central City area) on a south-facing slope in the ponderosa pine country. The property is now 260 acres. The elevation was between six and seven thousand feet, and the local creek (Cottonwood Creek in Gilpin County) drained into the north fork of Clear Creek, which flows eastward into Golden. On this creek was an old log cabin built by a settler in the early 1900s. It had two rooms that faced south, two windows in each room, and a wood stove. The original settler raised a few stock and somehow made a living there. He sold the land to another farmer, a Mr. Gullickson, who raised cows and had a garden about a mile upslope and who sold the property to me. It was perfect for a shack site. Luna came to visit, and we immediately began to plan a family party there to rebuild the tattered roof and start work on a real fireplace.

Log cabin at my Shack West, near Central City, Colorado. This is ponderosa pine country. Elevation about 6,000 ft. Old collapsed hay barn in foreground. Photo points northeast. Cabin was built by the first settler around 1910.

In 1970 we had a large family gathering at "Shack West." I prepared by hauling up lumber, tarpaper, nails, shingles, and gear so we could do serious roof-raising. That was really fun. A year or so later we had another family gathering to build the fireplace. To establish a firm foundation for a fireplace, Luna told me to dig a pit about three feet deep and fill it with cinder block and cement. The family joined in. The access road in was a mile-long mud drive dropping some seven hundred feet, partly dug out of the slope above the shack area, so one needs a four-wheel drive vehicle to get there.

I hired one of my young cousins, Jim Spring of Burlington, Iowa, to build a fence on the north margin of my property, which was useful for keeping out the neighbor's wandering stock. Once in a while thereafter my fence seemed to get cut, and we had to drive the cows out on foot. Not an easy job!

My family learned to enjoy my Shack West. Nina was living with me in Denver after her divorce from her first husband, and I reintroduced her to Charlie Bradley; Charlie then came down periodically from Bozeman to visit us and to help build the chimney at Shack West. I had built beds in the two rooms, and by then we could cook on the fireplace and on the wood stove. Once we had a good roof, good frame windows, an excellent fireplace, and a collection of tools, Carl came to visit with his new friend, Lynn Bradley (Charlie's niece). They courted at Shack West and, like Nina and Charlie, were later married at the Leopold Shack.

The main function of Shack West was to offer the land protection. When I first got the land there were signs on the fence posts in Spanish and English, "PELIGRO—Danger," announcing that poison pellets had previously been sprinkled (by plane) over the five hundred acres I had originally bought. The pellets were to poison predators (e.g., wolves, mountain lions, coyotes). I contacted Fish and Wildlife and let it be known that I wanted no such treatment on my land. At the beginning there had been no coyotes, few birds, and no rabbits, and I was concerned that my land had been badly degraded. The winter bird fauna was fairly diverse, as there were chickadees, crows, jays, and such. But summer fauna had been scant. After about two years I saw my first rabbit.

Shack West ultimately developed a rich local fauna, now replete with partridges, bears, mountain lions, bobcats, and, in the winter, elk. A golden eagle has maintained a nesting site at the lower margin of this property. One of the things one learns is that it is critical to fend off the developers who want to exploit the land next door and the miners who want to blast the rocks along the lower margin of the land for road mettle. Vigilance and friendly lawyers have helped protect the place; now a conservation easement is in place so the land at least cannot be developed. But watchfulness is needed.[11] The Shack West land is now protected by a conservation easement.

My heirs will be Nina's and Carl's grandchildren, some of whom live in the area. I was lucky to have had a good friend, Vim Wright, to help with the rebuilding, and now her son John sits on the management committee to help take care of the land. One of Caryl Leopold Smith's sons, Carl Smith of Fort Collins, Colorado, serves in that capacity too.

The Shack idea has been contagious in our family. Luna's daughter Madelyn and her family have established a Shack/farmhouse on about ten acres by a small wild fishing creek in the Driftless Area of southwestern Wisconsin. They are striving to establish a patch of original prairie. Madelyn and Claude's daughter, Clare, has inherited one of the cabins that Luna and students built on the slopes above Lake Fremont in Wyoming. Carl's daughter Susan Leopold Freeman and her husband, Scott Freeman, have established a Shack built on land that drains into Tarboo Bay in the Puget lowlands of Washington. They have gotten into the habit of planting conifers, thinning conifer stands and acquiring nearby land that they are protecting with easements. Their work in re-establishing meanders of the previously straightened creek has permitted salmon to move up the creek seeking gravel beds, which is exciting. Nina's daughter Trish Stevenson and her husband,

Gordon, acquired an old log farmhouse and about eleven acres of fertile valley land in the Driftless Area not far from Madison. They made the old structure comfortable by adding a huge wood stove and modern kitchen. They and their son, Bergere, are raising produce on the valley bottom for themselves as well as for sale in nearby markets.

Establishing a Shack has been a very positive process, not only for the Leopold siblings, but also now for the younger generations. We trace the Shack idea back to Dad and Mother, and their archery and hunting exercises and interest in land restoration. The lands that the young people of today are saving from development and the beauty of the land they and their children are actively protecting make it all worthwhile. Mother and Dad would have been proud, I am sure.

Ten

Epilogue: Family and Familiarity

"As many as possible should share in the ownership of the land and thus be bound to it by economic interest, by the investment of love and work, by family loyalty, by memory and by tradition."[1] This statement by Wendell Berry in his *Art of the Commonplace* says it all, in my view. Each of those elements can be so important in attending to one's land. The fact that the Shack land purchase during the Depression came at a reasonable price was indeed a fortunate happenstance, as it gave us a new opportunity. These eighty acres of failed Wisconsin farmland at the Leopold Shack that became Dad's experiment carried us all, a family of seven, into a communal work project. Wendell Berry's comment that "ancient wisdom... tells us that good work is our salvation and our joy"[2] was indeed our experience.

Aldo Leopold was an astute observer of nature and did not miss much. He traditionally carried a sharp pencil and a tiny notebook in his vest pocket where he noted pertinent developments and events that he wanted to register in the Shack journal or think about—things he was seeing that might be part of a new idea about the land. Some of what he saw would rub off on us, when he would turn to us and ask probing questions to explain it. One instance I particularly remember. We were standing by the river, when he pointed and said, "Baby, look at that island. Why do you suppose that the big cottonwoods are on the upstream end of that island and the little ones are growing at the bottom end?" I struggled with the issue until I remembered that the river was moving sand all the time to the downstream end of the island. That meant that the upstream end was *older* and the lower end was younger. Ha! That's the answer. In fact, theoretically one could date the upper and lower ends of the island using cottonwoods. A nice thought. Dad was a great storyteller. That made him a good teacher.

Mother was the glue that made everything fit nicely and work well; her family loyalty complemented Dad by providing the basic warm love that held our family together. It was Mother who was the crucial cultural element—in her case, Hispanic—in our family life, teaching us morals, kindness, social graces, and how to be happy. Her charm, her thoughtfulness, and her sense of humor were infectious and her talents impressive. Mother loved our experiences at the Shack just as much as the rest of us did. She was very positive about it. We all adored her.

The Shack projects were daunting but fun. The challenge of rehabilitating the little barn, using local materials, seemed natural. We could not afford much. But the place became a theater in which we could live simply, experience what was

Mother and Dad sitting on a "Leopold bench" near the woodpile in front of the Shack. Note the enormous oak section in front of them. Shack is in background.

growing around us, and enjoy all of it. Learning the flora and vegetation types—a continuing project—did not hurt us one bit. It drew our interest. We could imagine even on a small scale what the land had been like when the buffalo roamed here. There was plenty of time to wander, explore, to be alone, and to learn by doing. The nucleus for the pervasive harmony was of course our parents, who were the happiest married couple I ever knew. We were lucky. They treated us with love and respect.

It may seem odd that as children we always *wanted* to go along on Shack trips. It certainly was not mandatory. Carl upset his girlfriend when he announced he could not take her to the prom because he had to go plant trees during spring break at the Shack! He was talking to the girl he eventually married, Keena Rogers. We had a big project going at the Shack, which would be a great deal of fun, and we all wanted to be there. Many times we invited our friends to join our Shack trips, and that was a wonderful way to get to know people. We were fortunate. You find out a lot about people when you go camping with them.

To limit and restrict young people from a chance to experience native habitats or exciting places in nature is to rob them of something important and potentially wonderful. If we cannot teach our youth to love the land, then who, I ask, is going to defend it? Who will fight to protect it? If, as Loren Eiseley said, Nature is our hidden teacher, then we need contact with Nature to learn about her.[3] I am convinced that for Aldo Leopold and people like him, Nature was his teacher—and he paid close attention to her. He could ultimately see what were the best lessons from Nature.

Being free to explore and get into trouble at the Shack was a useful experience. We learn a lot from that too, such as by falling through the ice, and how to keep out of the way when the river ice collided noisily with the shoreline. Such a place is impressive for both its dangers and its opportunities. Exploring our woods with my new copy of *Spring Flora* and a hand lens, trying to identify my first plant to its Latin name.[4] That was a pleasant, quiet botanical exploration, a middle school experience that stuck. I learned that this flower, *Lysimachia quadrifolia*, has a special role in the bottomland woods. Another time, when my parents were napping and I was in our woods, I came upon a little fawn, a baby deer of rich brown color with white spots and big eyes, hiding in the bushes. I was careful not to frighten him or her, and backed silently away so his mother could find him again. I feel compassion for that little fawn, for that day. I was meeting someone who belonged here.

People who spend time close to the land may become observant and comfortable there. Maybe the word is "at home" on the land. It was certainly true of members of our family.

As we grew, we developed deep feeling for the Shack land. We sang together; some were songs about love, others about tragedy. Some of our songs in Spanish were about the land. For example, there was "Cuatro Milpas," about the little ranch now abandoned and lonely, "the place where I met my first love." There was "Mi Ranchito"—"Al pie de aquella montaña donde se oculta temprano el sol," about the sun rising every morning across the valley from behind that mountain. And there was "En las Playas del Mar" ("En las playas del mar varias conchas se ven"), about the lovely shells that come to the beach. Plaintive and gentle, the verses expressed a connection with the natural world. That "Playas del mar" was a song which Mother and her sisters sang together around Grandfather's piano in their youth in Santa Fe. This and many other rich songs became a part of our Shack culture. The Spanish language augmented our heritage. Singing was one of our outlets.

Life at the Shack somehow filled certain social and spiritual needs we had as a family. Our experience there was almost certainly the reason that each of us siblings turned to studies of nature and natural systems in the out-of-doors. Each of our careers turned to exploring an ecological system.[5] My brothers became almost as widely known in their sciences as Dad. Starker followed in Dad's footsteps as a wildlife ecologist and professor. Luna sought to solve problems involving rivers and erosion phenomena. Carl turned to plant ecology and then the mysteries of plant and seed physiology. He was a leading plant physiologist in his field, with an international reputation. His contributions in understanding plant reproduction in tropical rain forest are remarkable. Nina pursued wildlife ecology in the field with her first husband, and became a well-known and popular lecturer on the essence of the land ethic. In her talks Nina drew a lot of public attention across the United States to Dad's creative ideas about the land ethic.

As mentioned, I grew up as a field botanist learning about plants. Later as a research botanist I worked for the US Geological Survey, where I was charged with finding out if we can use fossil pollen grains to determine the age and environments of sedimentary rocks (it's the field of palynology, and yes we can). Later I taught forestry and botany at the University of Washington. It was interesting exploring past landscapes geologically using plant pollen and fern spores found in rocks and in the mud. It was an exciting way to reconstruct some of the glories of the past.

This book is about two things: familiarity with nature and togetherness. The Shack visits were experiences in simplicity, in getting along in the out-of-doors without a bunch of belongings. Recognizing land features became a habit: admiring special trees, certain visiting birds, and the call of the blue jay that announces to everyone that a stranger is walking in their woods. We loved showing all these things to our friends. The peent of the male woodcock that declares joy at the arrival of spring. The sounds of water rippling around snags in the river—"our river"! Here in this special place with its familiar setting a certain feeling of possessiveness, of belonging comes over you. You are a part of the community. Without realizing it you feel love and delight.

Part of that is connecting with one's family. We did so much together! I had the good fortune of being the youngest one, which gave me the maximum amount of time with my siblings and our parents in the Shack years.

In the Colorado foothills in the ponderosa pine zone I think I learned the hard way about feeling at home at my "Shack West." At first I felt alone and perhaps needy. That first summer I began to watch closely the birds that inhabited the bushes in front of my shack, and other fauna around me. One neighbor was a MacGillivray's warbler. Each spring and each year, for several years, I noticed that "he" nested in the same bush by the stream. I could watch and hear him working the area to catch bugs for his young ones. And I had a friend in "Squeaky the Wren," who sang at me through the log wall of my cabin every dawn in summer. I was amazed that this same individual wren with the uniquely squeaky voice came back from Central America every year for four years running! Then after a year I saw that first bunny. Certain golden-sided ground squirrels and striped chipmunks played in my front yard, digging happily. When in my third year I heard my first coyotes sing, that made the whole land experience just wonderful. Shack West had a new top predator! We saw coyotes on my meadow. With the winter birds that were so dear coming to eat grain at my window feeder, I was surrounded by friends.

The enrichment of the land community constituted what I call the "greening of Shack West" in Colorado: it was an area that under my protection was now freed of the terrible strychnine poison pellets that had wiped out the original coyotes, and doubtless many birds. My land was now free of grazing cows (except temporarily when the fence broke). I have been happy to know from the occasional tracks and exciting personal encounters that my land also includes a family of black bears and a family of mountain lions. A conservation easement now protects the area (260 acres).

At the Shack in Wisconsin, we felt complete familiarity with the local landscape, like the individual trees that we got to know well: the big cottonwoods on my island where Carl and I were always hoping an eagle would come and build a nest. During the winter after Dad passed away, I was feeling homesick as a graduate student in Berkeley, and I missed going to the Shack. Thinking about those trees, I sat down and wrote a song called "Cottonwood Tree." In the song I was imagining that I climbed such a tree (impossible, as they have few branches—and certainly not barefoot!). It was dream-world stuff. Music is an outlet that we turn to when necessary. The song is included in this essay.

To be out alone in nature is to be oneself. I wonder whether that experience is going to be a disappearing phenomenon, as people increasingly go in groups to visit nature, or don't go out at all. In our society today we are experiencing a kind of retreat from nature. Denying a child the freedom to wander and explore in the woods "because it might be dangerous" seems like a new wrong idea. What may be lost is a chance to develop a sense of community with the land, and an opportunity to know its beauty and diversity. A strong connection exists, of course, between honoring nature and developing ethical values. Dad wrote eloquent essays about it: "That land is a community is the basic concept of ecology, but that land is to be loved and

Cottonwood Tree

Estella Leopold, 1949

When I went out walking barefoot one day, I spied a cottonwood tree He was lone and straight and wonderf'ly old, He was tall and wild and free oh-oh, he was tall and wild and free

Well I stood right under that cottonwood tree
Looked up in his branches on high
And I heard that old south wind a blowin' his leaves
And I climbed that cottonwood tree, oh-oh
That big ole cottonwood tree

I asked my self how long it might be
Since he began to grow
And he answered me back in tones deep and low
Oh a hundred long years ago, oh-oh
A hundred long years ago

When I was a sapling young and green
I watched the geese in the night
And they flew over high and circled in low
There was dark ones and speckled and whites
Dark ones and speckled and whites

Now gone are the snow geese and gone are the pines
That around me used to grow
Well I'll go too some day just like I came
A hundred long years ago, oh-oh
A hundred long years ago

Repeat first verse

It was in 1949 when I was a grad student at UC Berkeley, homesick and missing the Shack trips, that I made up this song.

respected is an extension of ethics."[6] Wes Jackson was talking about this when he wrote in 1981, "Where there is alienation, [land] stewardship has no chance."[7]

Just what is it that makes one feel so warm, liberated, and free when one is out of doors? Familiarity. And, I would add, beauty. To revisit the very same land over and over in all the seasons, to watch the same plants come into bloom each spring, and to welcome them and the insects that pollinate them is a special privilege. Over time these plants, butterflies, and birds that appear together on a celestial schedule build our awareness of the interdependence of these species. One feels a sense of the ecological web of life. To watch and hear the migrants such as the elegant sandhill cranes or even the house wrens returning to one's land from someplace south each season is special, as they have each flown thousands of miles to return—to *our* farm! They become one's neighbors. These delightful avian fauna belong on this land, and the place is poorer without them. Naturalist Robert Pyle once wrote fondly about the migrant birds on his land, "If you wait long enough they will *all* come home."[8]

Dad wrote about nature in his most loving tones: "Our ability to perceive quality in nature begins, as in art, with the pretty."[9] I would add that to witness this special kind of natural beauty can set the stage for discerning poetry in the land. It can also be a basis for perceiving ethical values in one's life.

Aldo Leopold relaxing (*directly above*). This photo of was taken by Charles Bradley when he visited us at the Shack, perhaps about 1947. Photo of Mother courtesy of Sue Coleman Bennett (late 1960s).

Appendices: *Three Pet Stories*

THERE IS SOMETHING WONDERFUL ABOUT MAKING FRIENDS with an animal. Bringing in pet animals from the wild is not necessarily good for the individual pet, who may lose touch with its kin. But there are advantages to both individuals in the relationship too. My feeling is that it is good for young people to care for and get to know an animal over time. The experience gives one the opportunity to communicate, love, understand, and work with a member of the animal kingdom. The earliest pictures of Mother and Dad show them with a pet dog. Of course for Dad that was usually a hunting dog that would retrieve game birds shot in the field. Our family has always had pet dogs; we all loved them and considered them part of the family. But we also had other kinds of pets—hawks, other birds, and squirrels.

Because we had a love bond with our pets, we wanted to be kind and thoughtful to them, treat them in an ethical way, and protect them well. We talked to them. We learned to appreciate our animals, and we learned a lot from them. The communication was always fun, and important. Above all, to get to know an animal is to have the opportunity to be thoroughly enchanted by them, as I was, not only by our dogs but by the three crows whose stories are told below.

Sammy the Crow

When I was age seven or eight, a neighbor friend, Jimmy Telford, and I used to ride our bikes past the cemetery and into the oak woods around the reservoir at the westernmost edge of Madison.

It was perhaps the spring of 1934 or 1935. One day Jimmy climbed up an oak tree to check on the baby crows in a nest we'd been watching. He reported they were just tiny youngsters with eyes not open yet. This information gave us the idea that perhaps we could raise a young crow. Maybe each of us could raise a young crow as a pet. We came back two weeks later, and Jim climbed again. He reported that now the little crows had their feathers fledged out and eyes open. It seemed they were probably ready to leave the nest. Jimmy climbed down, holding a pillowslip containing two of the four baby crows from that nest. I called mine Sammy. Bringing Sammy home meant I had to refurbish an orange crate as a temporary nest. In those days orange crates were *the* thing to build with—they were wonderful. I put Sammy in our garage and fed him bread dipped in egg and small bits of raw hamburger, which he took willingly. He grew rapidly. Soon when I took him out to the backyard, he could flop around, trying his wings, and then rest in the sun.

Sammy quickly learned to fly around, mostly in the backyard. As he became more independent, he would decide not to return to his perch in the orange crate, but stay out in a tree somewhere. Then in the morning he would wake up early—about 5:00 a.m.—and want to be fed. Sammy's cacophony would awaken the entire family.

Nina and I used to laugh about the early morning scene out the back window: Dad would be in his pajamas and slippers in the garden with a shovel or garden fork digging worms for Sammy. I could just imagine Dad saying under his breath, "Well goddamn! Now be quiet—here's another worm!"

As summer developed, Sammy was able to fly to the top of the neighbor's house, perch on the chimney, and at 5:00 a.m. raise his voice, "Caw Caw Caw Cawwww," right down the chimney. Soon after that, Mother received a call from Mrs. Montgomery, a dear next-door neighbor, saying, "Do you think young Estella could please take that crow inside at night?" Mother passed that on to me. After that, Sammy was confined to the garage orange crate perch at night after being fed.

Sammy was a real prankster. It turned out that this was the summer when Mother had arranged to send me to the Joy Camp in northern Wisconsin. I wrote ahead and asked Miss Joy if I could bring my pet bird up to camp with me, because I could not come otherwise. The answer was, "Yes, bring your pet, and you will take care of it."

So I boarded the train in Madison, carrying my crow in the orange crate that had the perch in it and a bowl of water. The bird was assigned to the baggage car. When we got to Joy Camp several hours later, I picked up the bird, my new guitar, and my luggage, and was escorted to the cabin where six of us girls would stay. At first the crow was a sensation. He liked all the attention and would fly around the camp, inspecting the outsides of the cabins. He especially liked to wait for us when we finished lunch in the mess hall, where we would buy stamps and writing materials. The girls would fondly lean over to talk with Sammy as he sat by the trail. Sammy would scrunch down low, bat his eyes, and look disarmingly up at the admiring campers. They would gesture to Sammy with a postcard or a book of stamps. Every once in a while Sammy would grab the stamps and fly away with them, to the distraction of the owner. Pretty soon we discovered that it was Sammy who had collected some of the colorful toothbrushes we kept on the outside of the cabins with a cup to brush our teeth. After our toothbrushes began to disappear, we found a whole pile of these and a few stamps and postcards in the woods—Sammy's stash! Everyone laughed, but not everyone was entirely appreciative.

Sammy liked to watch the girls get on horseback and go for a ride in the woods. He would follow along, staying above in the trees, enjoying the company.

At the end of the season, the counselors had arranged for us to put on a play to entertain the parents who had come to collect their daughters after the four-week term. The theater was a natural amphitheater, with a flat area that was the stage, and a small half-circle ridge around the flat area where the audience could sit and watch the play. On this occasion, the weather was fine. The parents and some of us campers were sitting in rows on the embankment, watching the play. I could see Sammy up on the far side of the amphitheater at the top of the trees also watching the play. Then I saw him set his wings, jump off the treetop, and swoop down over the girls who were acting. Then he dive-bombed the audience, who all tipped backward with a great cry of dismay. The unwanted surprise attack from a strange black bird abruptly ended that event. There were loud ohhh's coming from the audience. There was nothing I could do, as Sammy disappeared in the woods. The counselors moved to center stage and announced to the audience that perhaps this was a good time to go to the mess hall and have lemonade.

That evening after supper, the counselors told me that Miss Joy wanted to see me in her office. I meekly went to the office, expecting a scolding. Miss Joy said, "Estella, I have written this telegram to your mother, and I want you to read it." The message said: "Dear Mrs. Leopold, I am sending Estella's crow to you in Madison by train COD tomorrow morning. Sammy has enjoyed being here, but his tenure is over. Miss Joy."

The next morning I equipped Sammy's orange crate and perch with a bowl of water and a packet of food with directions to the train man to feed the bird some hamburger for the trip. We loaded him onto the baggage car, and he was gone.

Mother dutifully met the train from northern Wisconsin and returned Sammy to the backyard at home. I know it must have been a burden to her to have to feed the crow and to try to keep track of him. Each night she had to get him into the garage—no easy task. Two weeks later, when I returned from camp, Mother picked me up at the train station, and I asked about Sammy. She told me, "You know it is not easy to keep track of Sammy, Estella. He is pretty independent. We are not quite sure where he is right now. Did you have a nice time at camp, dear?"

In the next few days, I rode my bike around looking for Sammy. Someone told me they had a new crow in the Madison Zoo. So I immediately went to the zoo on my bike. Sure enough, there was a mature crow in the zoo behind bars! I explained to the zookeeper, "Oh, that is Sammy, my pet crow! He recognizes me! See! Look. He talks to me. You need to let me have him back." But the zookeeper was unimpressed. "No," he said, "this is not your crow. It is the zoo's crow; you must be mistaken."

Well, of course, I was heartbroken, as it was so very obvious that this was Sammy, my crow. Why couldn't the zookeeper see that? Why would he not believe me? I did go visit Sammy from time to time, but was not able to free him. I told Mother and Dad that the zookeeper did not believe that this was our family pet. Dad said to me, "You know, Baby, your mother did her best to take care of him. We can thank her for that."

Pedro Visits a Tenth-Grade French Class

When school resumed one fall, I was attending tenth grade at West High. Pedro, my second pet crow, seemed to get lonely and wanted attention during the day. Little did I know that this new crow would change my life, and my grades! Having seen me take off for the West High School across the street each morning, he began to look around and see what was going on in that building.

One day he discovered me through second-floor windows in French class. To the class's amusement he began to come visit. He would walk back and forth on the cement windowsill making little cooing noises. I must admit I was pleased to see him but very worried. The class kept looking at Pedro instead of at the board where the French teacher was writing and trying to teach. Our teacher was naturally upset at this disturbance, and, figuring out that this was my crow, he handed me a note and told me to report to my homeroom teacher. I was really scared, because my homeroom teacher was a frightening older lady whose hand shook slightly as she read the teacher's note. It read "It's that crow again! Signed Mr. X." Miss Wilson looked at me and said to please sit in the back and wait till her class was finished. I was terribly impressed as members of her senior class stood one at a time to translate Cicero and read it in Latin and in English. Wow, were they ever good! One of these was a star student named Ed Severinghaus, whose father also taught at the university. I felt like an interloper! Being younger, I felt quite inadequate and frightened. When the students left Miss Wilson called me to her desk and said, "You have to do something about your crow! It is very disturbing to everyone. Now you need to do something about your grades. I have had your sister Nina and brother Carl, and they both did excellently in my classes. You are getting poor grades, and I want you to turn over a new leaf and get on the honor roll right away! Do you hear me?" She pointed her finger at me and shook a bit. Miss Wilson gave me a fierce look. "There is no excuse for your getting poor grades. Now *go*. You are excused!"

I went. I was quite terrified. Miss Wilson was a very strict teacher. And I did begin to work hard, and did get on the honor roll right away. I also kept Pedro in his box on the mornings I had French class. That helped some. Pedro was glad when school was out for the summer, however.

Later in the summer Dad was going to Riley, where they had a hunting cooperative, to talk with his farmer friend Mr. Reuben Paulson. Did I want to come along? I did, and I took Pedro with us. As we stood in Mr. Paulson's front yard, a great flock of crows was flying over, calling and cawing. Pedro seemed very excited. I lifted him up and gave him a little push, whereupon Pedro took off, flapping upward and joining the great flock of crows. He was on his way. I watched as the great flock disappeared, heading off somewhere. I felt happy for him. He deserved to be with his own kind. What a good experience it had been to have a friend like Pedro. We liked each other a lot. I was sorry to see him leave, but of course respected his need to be with other crows.

Fluminea, the Manitoba Crow

In the summer of 1947 my sister Nina and her husband Bill Elder went up to Manitoba to work at the famous Delta Waterfowl Research Station. My father in his role as chair of a technical committee of the American Wildlife Institute had helped establish this research station on the marshy shores of Lake Winnipeg. Just east of the station was the famous Clandeboye Bay, where sandhill cranes liked to breed. I was in college in Madison studying botany, and Nina and Bill asked me to come visit them for a period during the summer. The Delta Station was run by another of Dad's graduate students, the famous artist Hans Albert Hochbaum. There were other researchers working there, such as Lyle Sowls, another of Dad's graduate students, and Lyle's wife, Grace, who had a new baby. Interesting researchers like Frosty Anderson, a longtime Audubon employee, were there—all these people and more, such as Hans's colleague Peter Ward, also a famous waterfowl artist.

The station buildings were frame houses spread out along the south shore of Lake Winnipeg near the town of Portage La Prairie. Nina and Bill had a house just west of Al Hochbaum's. There was room for two of Bill's grad students, Sandy and

Bob, who were developing their own research and helping Bill with his studies of lead poisoning of waterfowl. I felt fortunate to be there too, and enjoyed seeing new country and being helpful to Nina and Bill.

Bill and Nina picked me up in Winnipeg at the train station, and we drove across the flat country which was the old bed of glacial Lake Agassiz, to Portage La Prairie and the Delta Waterfowl Station. A population of varied water birds, marsh wrens, western grebes, redheads, shadpokes, and such, made it a very lively interesting place.

Right away I met some of the neighbors, including a young boy about twelve years old named Andre La Fleche, who spoke fluent French and some English. One of the researchers had found a young crow and was trying to decide what to do with it, and Andre and I volunteered to feed it and take care of it while I was there. Andre was immediately a help in caring for this crow. We named the crow Fluminea, for the Latin name of the tall marsh grass that prevailed all over these flat lands. We built Fluminea a box with a perch in it, and fed her regularly using hamburger, bread dipped in egg, and such. As a matter of experience we had her sleep in the box indoors for her safety. But during the day Fluminea had the run of the research station, visiting the neighbors, flying about the neighborhood, teasing the dogs, and having fun. I would feed her at night, at which time I could place her into her box on a perch for the night.

The major researchers at Delta were all huntsmen and fond of shooting waterfowl in season. They each had a black Labrador retriever as a hunting dog. To take care of these three Al and Peter had built a large dog house, about seven feet square, with a low roof and a wide door. It had a waterproof roof, and blankets for the dogs to lie on.

Fluminea took a great interest in the black dogs. She began by cawing at them, walking near them, and teasing them by flying low ahead of them. They were excited and raced after her, till she crossed over a wire fence and they had to stop while the crow flew bravely on. She was a very talkative crow, and she liked to collect trinkets around the neighborhood.

One day we were sitting at lunch at Nina and Bill's with the student helpers, Sandy and Bob, when we suddenly heard a splashing noise above us. I quickly realized that Fluminea had tried to get a drink out of the huge water-catchment container on the roof and was splashing around on its surface. I leapt up, went up the ladder and, climbed up to the rim of the water tank, and peered down. There she was, splashing around with her wings. I could not reach her, so I hopped over the edge and hung on to the rim to let her get on my foot so she would not drown. Bob was right behind me, bless him, and pulled me up out of the tank with the crow on my foot, till we had her back on the roof, and me too. What a shock!

We patted down Fluminea with a towel, fed her, and helped her recover from this trauma. Next day she was just fine. About a week later there came in a huge thunderstorm, with lightning and wind and torrents of rain. I realized that Fluminea was not to be seen and was out there somewhere in the storm. I ran outside calling her over and over, "Come here, Fluminea, Come here. Caw Caw Caw!" I could not find her anywhere. The rain kept pelting down, and the thunder was frightening. I was getting soaked. She was not in any of her usual places. I called and called, walking up and down between the buildings, but no Fluminea. It was still raining hard. On returning to our house, I paused, leaned down, and looked into the dog house. Well! There were three black Labs lying together out of the rain, and there was Fluminea, walking up and down between them! I could hardly believe my eyes! "Fluminea," I shouted! "*There* you are! What are you doing in there with those dogs? Come on, I'll take you in!" She nicely let me pick her up, while the dogs quietly looked on—I put her under my arm and took her in to our place. I patted Fluminea down with a towel, and put her on her perch in her box. What a rainstorm! But she had been perfectly at home with those big black dogs. Incredible! She might have just stayed with the Labs awaiting the end of the storm!

Toward the end of the summer, Andre La Fleche took care of Fluminea when I went on field trips to see more of the country. Nina and I went on a ten-day canoe trip with friends, and Fluminea stayed home at Delta under the care of Andre, who liked Fluminea very much. He cuddled her and said nice things to her in French. At the end of the summer, she stayed on with him. One of Dad's graduate students published a picture of Mother with a crow in Madison, claimed to be Fluminea (see *The Professor**), But the photo was of a previous crow, as Fluminea remained in Manitoba with Andre La Fleche at the Delta Research Station when I went home. She was a friendly and amusing character, and we all enjoyed her. What a nice crow she was!

Where Did They Come From?

A few words about the families of Aldo Leopold and Estella Luna Bergere.

Aldo Leopold's Family

To appreciate the background of this special and honorable man, we need to look back to Aldo's parents and grandparents.[1]

Aldo Leopold's maternal grandfather, Charles Starker, of Stuttgart, Germany, attended Beale School and Polytechnical School studying architecture and landscape engineering. He worked four years for the Bavarian government supervising the erection of bridges and locks on the Danube. He decided to immigrate to the United States in 1848. Mr. Starker worked for Gray and Co., hide and leather merchants, in Buffalo (for eight dollars a month). Then in Chicago he began work with the architect T. Knudson. He was asked to design a house for Territorial Governor Grimes in Burlington, Iowa, and he fell in love with Burlington. The tall bluffs and view of the huge Mississippi River reminded him of his homeland and the Danube River. He relocated to Burlington and entered the mercantile business there. Charles Starker married Marie Runge of Burlington. They had two children, including Clara Starker, Aldo's mother.

During his early years in Burlington, Charles Starker chaired the committee that built the Opera House; laid out and designed the cemetery, as well as Snake Alley downtown; designed Crapo Park; established the Iowa State Savings Bank in 1874; and served in several other capacities in Burlington. In the 1870s he returned to Germany several times, and convinced about three hundred families to immigrate to the Burlington area. On one trip, he took his young daughter Clara and wife Marie with him for a visit with their relatives. Clara was particularly fond of the grand opera experiences she had there.

In 1872, Charles Starker moved his family to the spacious house now called Starker House on Prospect Hill in Burlington overlooking the river. It was and is a beautiful landscaped green area, a "bird's paradise." The newspaper of the day, the *Burlington Hawkeye*, said Starker made "an effort to preserve some of the greatest gifts of nature...birds that were daily visitors, returning to this idyllic spot with the seasons." Charles Starker was known for generally favoring more naturalistic landscape arrangements. He built the Bluff Trail above the river, which many have enjoyed since.

Charles Starker's daughter Clara married a local businessman (her first cousin) and outdoor enthusiast, Carl Leopold, in 1885. Carl Leopold was a Burlington native. His father, Charles J.J. Leopold, born in Hanover in 1809, had entered the University of Berlin, and then came to the New World in 1834. He settled in Saint Louis, where he worked at a dairy. He married Thusnelda Runge and moved to Liberty, Missouri, to start a rope-making business. Eventually they moved to Burlington, Iowa.

In 1888, Carl Leopold formed the Rand-Leopold Desk Company with C. W. Rand. They took over the Northwest Furniture Company, which manufactured revolving bookcases and wall desks. They started producing rolltop desks (under the motto "Built on honor to endure"), an item that became highly prized. This became the Leopold Desk Company in 1900. They were shipping furniture coast to coast after about 1900. These rolltop desks apparently caught the interest of businessmen who wanted to lock up their desks at night with a key.

Under Carl Leopold, and using skilled craftsmen, the Leopold Desk Company flourished. Their desks were of native maple, oak, and walnut or cherry. They sent their wood, cut to board lumber to, Herkimer, New York, for processing. Once there, the wood was densified so that the finished desk was dent-proof and strong, and also heavy. Then the wood was returned to Burlington for desk production. Before the time of the unions, Carl Leopold was rewarding his employees by giving them stock in the company, which created a very loyal crew of skilled labor for the Leopold Desk Company.

Like his father-in-law, Carl Leopold was active in community affairs, but most especially he was an outdoorsman. He loved to hunt, but he shot only what his family could eat. Son Fredric Leopold wrote that Carl Leopold took a sensitive approach to hunting. To him hunting wild game was a special privilege, "and game was always served sparingly." Carl stopped spring shooting back before son Fredric was born, apparently calling it unethical. The children caught on and adopted the idea.[2]

Carl Leopold made sure that his four children became very familiar with the native plants and animals of the region. On occasion he would drive the family out on the Iowa prairie where they could appreciate the native grasses and prairie chickens. Bounty included passenger pigeons, bobwhite quail, woodcock, woodchuck, and rabbit.

Carl's eldest son, Aldo, attended Lawrenceville School in New Jersey, and later earned a master's degree in forestry at Yale (1905). From there, Aldo embarked on his career in the US Forest Service and a long tenure as a field ecologist and professor at the University of Wisconsin.

Dad was a kind and loving father. He was always well dressed. He was pretty serious around Madison, where he was serving as a newly minted professor. But up at the Shack or on camping trips he was jubilant and happy and such fun! He loved showing us the outdoors, and would typically ask us if we had noticed this or that in the field. It was fun. He loved to tell stories and recount the various hunting trips he would take, to Mexico or to Missouri quail-hunting with his friends. Dad was a sharp observer, and had a way of putting together events that had happened in the field—he was a kind of astute field detective. Personally he was always polite and looked to find the best in every one he knew. As my brother Luna wrote, "Never would he talk down to a person and he treated people in menial positions with the same consideration and courtesy as he would the most exalted." His great love for his wife was very apparent. To him Estella Bergere was the most important part of his life. They were simply wonderful together.

Estella Bergere's Family

My mother's family, the Lunas and the Oteros, has a history that go back three hundred years: there were Lunas that sailed to Mexico with Hernando Cortez in the 1500s and on several voyages after that.[3]

Jose Enrique Luna (born 1771) had come up the trail along the Rio Grande from Mexico in the 1800s driving sheep and settled in what is now Los Lunas (named after his ranch site) on the San Clemente Grant, which he had purchased. His son Antonio Luna and wife Isabel (de Baca) became the parents of my grandmother Eloisa Luna and her brothers, who were much involved with politics during statehood for New Mexico.

An interesting story from the family tells that the original ranch house for Antonio Jose Luna and Isabel Baca was along the Rio Grande in the location of Los Lunas. The D and RG Railroad was very anxious to build a track line near the river to serve the township of Belén and approached Isabel and Don Antonio Jose Luna about selling their house site to give right of way for the railroad. Isabel put her foot down and said no, she was not interested. The railroad came back again and again asking for the right of way. Isabel finally succumbed when the railroad built a house to her specifications where she wanted in return for the right of way. Well, the railroad built the beautiful Luna Mansion in Los Lunas, and Belén became the railroad center of the state. The house is a mansion on a site with a fine *acequia madre* (canal) for irrigation and some large cottonwood trees that are there today. The lot ultimately became a vast garden and orchard. Isabel and Antonio Jose Luna left the mansion to their son Tranquilino, who was the elected delegate to the Congress from the territory. The Luna Mansion is still operating as an elegant Hispanic restaurant today.

Antonio Jose Luna and Antonio Jose Otero of Valencia were involved in a very bold venture in about the 1850s. They drove their sheep to California and sold them at a premium price during the gold rush: about ten to fifteen dollars a head, instead of at a local New Mexico value of fifty cents per head. Clearly that was no small feat, considering the immense distances and desert conditions, not to mention the danger from Indian attacks. *El Rio Abajo*, a book by Gilberto Espinosa and Tibo J. Chavez, estimated that the Luna and Otero families drove in one drive about fifty thousand sheep to California. These families quickly became "the big *ricos*" and political leaders in Valencia County at that time. When Salomon Luna married Adelida Otero of Valencia County, and when Salomon's sister Eloisa married Manuel B. Otero (1879), the Luna and Otero families became the most powerful and influential of the times. In the late 1800s my grandmother Eloisa Luna had been declared the most beautiful young woman in the territory, according to one press account. Her first husband, Manuel B. Otero, was a rich young sheep rancher from a prestigious family, the Otero's. After they were married their mutual holdings made them the largest sheep ranchers in New Mexico.[4]

My grandmother Eloisa Luna Otero was widowed at age twenty when her young husband, Manuel (age twenty-three), was killed by Joel Whitney on the Estancia ranch in 1883. Eloisa was pregnant and already had two children: Eduardo Otero and Nina Otero, who helped their mother and Uncle Salomon continue to run the ranch. Later, son Manuel was born. Nina Otero became a well-known person in the community later in the suffrage movement.

As a well-to-do young widow, Eloisa was soon courted by other gentlemen. One of them, Alfred Bergere, had met her earlier when he came through Santa Fe as a merchandiser. When he heard that this lovely person was widowed, he went to Santa Fe to court her. As a semiprofessional pianist, Alfred soon won her heart, and they were married in or about 1885. They lived in Los Lunas, where my mother was born, until Governor Miguel gave Alfred Bergere a job in Santa Fe. They moved into the "big house" at 135 Grant Avenue where my mother grew up. Mother had eight sisters and three brothers. That large adobe house is now the headquarters for the Georgia O'Keefe Foundation. After her schooling, Mother was sent to Saint Louis to enter Marysville College of the Sacred Heart, a kind of finishing school. Mother returned to Santa Fe and was teaching school when she and her sister Anita went to Albuquerque to a dance. There she met Aldo Leopold.

Our dear mother was such a glorious person! First, she was always beautifully groomed with black wavy hair tied in a bun on the back of her head, as her sisters wore it in New Mexico. She was handsome and dark-skinned, with a lovely speaking voice. One time we had a visit from a famous British ecologist, Fraser Darling. He described my Mother this way: "Mrs. Leopold carries in her presence all that Spain might have given to the world, graciousness, gravity, pride, devoutness, perfection of hospitality, and a quite amazing gaiety, the impact of which in its sudden unexpectedness makes her one of the most beloved of women. As for me I loved hearing her speak, for she used a slow, beautiful English purer than that which the English speak."[5]

Mother was very musical and had taken piano lessons from her father, Alfred. She sang Spanish songs to us and told us wonderful stories. She was fluently bilingual and had a great sense of fun and happiness. Mother was a great cook, making delicious whole wheat bread for our table every week and preparing great meals. She was kind and generous and had many fine friends, including some faculty wives and social leaders—even forming a reading club with them. The eight of them would get together about every two weeks and serve tea and little cucumber sandwiches, and they would take turns reading out loud from current books of interest. As small children we always loved to hear them telling stories and laughing together. They were warm good friends and had a fine time together.

Mother was charming and fully supportive, always delighted to see Dad when he came home for lunch and after work each day. Mother was the disciplinarian of the family and watched out over us making sure we grew up with ethical standards. From her we learned the politeness that was the behavior of the times. She was beloved by all who knew her. Her support and affection for Dad was unflagging and obvious to everyone who knew her. I never heard a harsh word between them. That was just the way it was!

Notes and Sources

Preface

1. Aldo Leopold, *A Sand County Almanac, and Sketches Here and There* (New York: Oxford University Press, 1949).

Chapter One

1. Aldo Leopold, *The River of the Mother of God, and Other Essays by Aldo Leopold*, ed. Susan L. Flader and J. Baird Callicott (Madison: University of Wisconsin Press, 1991), 211
2. A. Leopold, *Sand County Almanac*, 47.
3. Stephen A. Laubach, *Living a Land Ethic: A History of Cooperative Conservation on the Leopold Memorial Reserve* (Madison: University of Wisconsin Press, 2014); census data.
4. Laubach, *Living a Land Ethic*, 31.
5. Laubach, *Living a Land Ethic*, 33.
6. Laubach, *Living a Land Ethic*, 33.
7. A. Carl Leopold, personal journal, vol. 1, 1934–1936, archives, Aldo Leopold Center.
8. A. Leopold, *Sand County Almanac*, 25 ("Come High Water").
9. A. Carl Leopold, personal journal, vol. 1, 1934–1936, archives, Aldo Leopold Center.
10. A. Leopold, *River of the Mother of God*, 343 ("The Ecological Conscience").
11. Asa Gray, Benjamin Lincoln Robinson, and Merritt Lyndon Fernald, *Gray's New Manual of Botany: A Handbook of the Flowering Plants and Ferns of the Central and Northeastern United States and Adjacent Canada*, 7th ed. (New York: American Book Company, 1908); Norman Carter Fassett, *Spring Flora of Wisconsin*, ed. M. S. Bergseng (Madison: University of Wisconsin Press, 1967).
12. Curt Meine, *Aldo Leopold, His Life and Work* (Madison: University of Wisconsin Press, 1988), 364.

13. Aldo Leopold, *Round River*, ed. Luna Leopold (New York: Oxford University Press, 1953), 127 ("The Deer Swath").
14. Benjamin Franklin, "Description of a New Stove for Burning of Pitcoal, and Consuming All Its Smoke," *Transactions of the American Philosophical Society* 2 (1786): 57–74.

Chapter Two

1. A. Leopold, *Round River*, 127 ("Snow Tracks").
2. A. Leopold, *Sand County Almanac*, 11 ("The Good Oak").
3. A. Leopold, *Sand County Almanac*, 87 ("65290").
4. A. Leopold, *Sand County Almanac*, 25 ("Come High Water").
5. A. Leopold, *Sand County Almanac*, 6.
6. John Muir, *The Story of My Boyhood and Youth* (Boston: Houghton Mifflin, 1913).
7. A. Leopold, *Sand County Almanac*, 6.

Chapter Three

1. A. Leopold, *Sand County Almanac*, 25 ("Come High Water").
2. A. Leopold, *Sand County Almanac*, viii.
3. A. Leopold, *Sand County Almanac*, 19.
4. A. Leopold, *Sand County Almanac*, 22.
5. Robert Michael Pyle, *Wintergreen* (Seattle: Sasquatch, 2001), 115.
6. Frances Hamerstrom, *Harrier, Hawk of the Marshes: The Hawk That Is Ruled by a Mouse*, Smithsonian Nature Series 6 (Washington, DC: Smithsonian Institution Press, 1986).

Chapter Four

1. Aldo Leopold and Alfred E. Eynon, "Avian Daybreak and Evening Song in Relation to Time and Light Intensity," *Condor* 63 (1961): 269–93.
2. A. Leopold, *Sand County Almanac*, 51 ("The Green Pasture").
3. Aldo Leopold, *A Sand County Almanac, and Other Writings on Ecology and Conservation*, ed. Curt Meine (New York: Library of America, 2013), 683.
4. Marjorie Green Winkler, "Effect of Climate on Development of Two *Sphagnum* Bogs in South-Central Wisconsin," *Ecology* 69, no. 4 (1988), 1032–43.
5. Winkler, "Effect of Climate."

Chapter Five

1. A. Leopold, *Round River*, 6 ("Man's Leisure Time").
2. A. Leopold, *Round River*, 7 ("Man's Leisure Time").
3. A. Leopold, *Sand County Almanac* (2013 Meine ed.), 683.
4. A. Leopold, *Sand County Almanac* (2013 Meine ed.), 683.
5. A. Leopold, *Round River*, 106 ("Gus's Last Hunt").

Chapter Six

1. A. Leopold, *Round River*, 7 ("A Man's Leisure Time").
   According to Mark Hicks of Ozark, MO, who helped me track down the make of the bench, "The face vise was made in Chicago, by the Abernathy Vise and Tool Company, which leads me to believe that the bench could have been made by the Christiansen Workbench Company. This company originally produced benches with only wooden screw vises, but switched to metal Abernathy vises after the Christiansens acquired the company. The vise could also have been salvaged from another bench and added to this one by your father. Someone would need to take a very close look at the collar of the wooden screw vise to see if they can find the Christiansen name."
2. Aldo Leopold, *Glues for Wood in Archery*, US Forest Service Products Laboratory, Technical Note no. 226.
3. "Skill in the Sport of Robin Hood Achievement of the Aldo Leopold Family," *Capital Times* (Madison, WI), November 23, 1930, 16.
4. *Capital Times*, November 23, 1930, 16.
5. *Capital Times*, November 23, 1930, 16.
6. *Capital Times*, November 23, 1930, 16.
7. "Santa Fe Girl Wins National Archery Record," *Chicago Tribune*, August 24, 1930.
8. "Mrs. Aldo Leopold Is Archery Contest Victor," *Capital Times*, July 6, 1930, 8.
9. "Announcement of Hunt," *Racine Journal-Times*, August 31, 1934, 12.
10. Meine, *Aldo Leopold*, 340.
11. Robert Wegner, *Deer and Deer Hunting: The Serious Hunter's Guide* (Mechanicsburg, PA: Stackpole, 1992), 54.
12. Luna Leopold, unpublished journal of 1934, housed at American Philosophical Society Library, Philadelphia.
13. L. Leopold journal.

14. "25 to Hunt Deer with Bow, Arrow," *Milwaukee Journal*, November 22, 1934.
15. Meine, *Aldo Leopold*, 683.

Chapter Seven

1. A. Leopold, *Round River*, 146 ("Outstanding Discovery").
2. Mary Austin, *Land of Little Rain* (Boston: Houghton Mifflin, 1903), 88.
3. Robert H, Dott, Jr., and John W. Attig, *Roadside Geology of Wisconsin* (Missoula, MT: Mountain Press, 2004).
4. Lee Clayton, John W. Attig, David M. Mickelson, and Mark D. Johnson, *Glaciation of Wisconsin* (Wisconsin Geological and Natural History Survey, Educational Series 36, 1992).
5. Lee Clayton, John W. Attig, David M. Mickelson, Mark D. Johnson, and Kent M. Syverson, *Glaciation of Wisconsin*, 4th ed. (Wisconsin Geological and Natural History Survey, Education Series 36, 2011).
6. Winkler, "Effect of Climate Change."
7. R. O. Kapp, D. L. Cleary, G. G. Snyder and D. C. Fisher, "Vegetational and Climatic History of the Crystal Lake Area and the Eldridge Mastodont Site, Montcalm County, Michigan," *American Midland Naturalist* 123, no. 1 (1990): 47–63.
8. Kapp et al., "Vegetational and Climatic History."
9. John W. Attig, Lee Clayton, Kenneth I. Lange, and Louis J. Maher, *The Ice Age Geology of Devil's Lake State Park* (Wisconsin Geological and Natural History Survey, Educational Series no. 35, 1990).
10. Winkler, "Effect of Climate Change."
11. Muir, *Story of My Boyhood*.
12. John H. McAndrews, "Postglacial History of Prairie, Savanna, and Forest in Northwestern Minnesota," *Memoirs of the Torrey Botanical Club* 22 (1966): 1–72.
13. Timothy E. Crews and Philip C. Brookes, "Changes in Soil Phosphorus Forms through Time in Perennial Versus Annual Agroecosystems," *Agriculture, Ecosystems and Environment* 184 (2014): 168–81.
14. Aldo Leopold and Sara Elizabeth Jones, "A Phenological Record for Sauk and Dane Counties, Wisconsin, 1935–1945," *Ecological Monographs* 17, no. 1 (1947): 81–122.
15. Patricia Klindienst, *The Earth Knows My Name: Food, Culture, and Sustainability in the Gardens of Ethnic Americans* (Boston: Beacon, 2008).

16. A. Leopold, *Sand County Almanac*, 26 ("Draba").
17. A. Leopold, *Sand County Almanac*, 44 ("Prairie Birthday").
18. E. J. Dyksterhuis, "The Vegetation of the Fort Worth Prairie," *Ecological Monographs* 16 (1946): 1–29.
19. Shoshana K. Mertens, "Edaphic Variables and the Distribution of Prairie Forms (Wisconsin)," *Restoration and Management Notes* 8, no. 2 (1990): 99–100.
20. A. Carl Leopold, quoted in Thomas Tanner, *Aldo Leopold: The Man and His Legacy* (Ankeny, IA: Soil Conservation Society of America, 1987), 167.
21. F. Fraser Darling, *Pelican in the Wilderness: A Naturalist's Odyssey in North America* (London: Allen & Unwin, 1956), 66.

Chapter Eight

1. Laubach, *Living a Land Ethic*, 20.
2. Laubach, *Living a Land Ethic*, 41.
3. Laubach, *Living a Land Ethic*, 50.
4. Muir, *Story of My Boyhood*.
5. N. L. Bradley, "A Sand County Restoration," *Restoration and Management Notes* 5, no. 2 (1987): 77–79.
6. C. C. Bradley, unpublished, "The Bradley Study Center Log"; Charlie Bradley's journals, as unpaid codirector of research; reports; all in the Aldo Leopold Foundation archives.
7. Konrad Liegel, "The Pre-European Settlement Vegetation of the Aldo Leopold Memorial Reserve," *Transactions of the Wisconsin Academy of Sciences, Arts, and Letters* 70 (1982): 13–26; Konrad Liegel, "Land Use and Vegetational Change on the Aldo Leopold Memorial Reserve," *Transactions of the Wisconsin Academy of Sciences, Arts, and Letters* 76 (1988): 47–68.
8. Henry T. Lewis, "Indian Fires of Spring," *Natural History* 89, no. 1 (1980): 76–83.
9. N. L. Bradley, "Sand County Restoration."
10. Crews and Brooks, "Changes in Soil Phosphorus Forms"; S. W. Culman, S. T. DuPont, J. D. Glover, D. H. Buckley, G. W. Fick, H. Ferris, and T. E. Crews, "Long-Term Impacts of High-Input Annual Cropping and Unfertilized Perennial Grass Production on Soil Properties and Belowground Food Webs in Kansas, USA," *Agriculture, Ecosystems and Environment* 137, nos. 1–2 (2010): 13–24.
11. Lewis, "Indian Fires of Spring."

12. Liegel, "Pre-European Settlement Vegetation."
13. Liegel, "Land Use and Vegetational Change."
14. A. Leopold, *Sand County Almanac*, 95 ("Marshland Elegy").
15. Nina L. Bradley, A. Carl Leopold, John Ross, and Wellington Huffaker, "Phenological Changes Reflect Climate Change in Wisconsin," *Proceedings of the National Academy of Sciences* 96, no. 17 (1999): 9701–4.
16. David Alan Weinstein (n.d.), "Top Ten Responders in Phenology at Leopold Preserve, Wisconsin, USA" [PowerPoint Slide], Cornell University.
17. Leopold and Jones, "Phenological Record for Sauk and Dane Counties."
18. The Wisconsin DNR has been publishing a calendar Division of Cooperative Extension of the University of Wisconsin-Extension. 2015. Wildlife Phenology Calendar. Cooperative Extension Publications, Madison, WI. http://learningstore.uwex.edu/Wisconsin-Wildlife-Phenology-Calendar-P1382.aspx.
19. Steve Swenson, *My Healthy Woods* (Baraboo, WI: Leopold Foundation, 2013).

Chapter Nine

1. A. Leopold, *Sand County Almanac*, 218.
2. C. C. Bradley, Personal journals (unpublished).
3. Crews and Brooks, "Changes in Soil Phosphorus Forms."
4. J. H. Bock and C. E. Bock, "Natural Reforestation in the Northern Sierra Nevada Donner Ridge Burn," *Proceedings of the Annual Tall Timbers Fire Ecology Conference* (1969): 119–26; C. E. Bock and J. F. Lynch, "Breeding Bird Populations of Burned and Unburned Conifer Forest in the Sierra Nevada," *Condor* 72 (1970): 182–89; J. H. Bock, C. E. Bock, and V. M. Hawthorne, "Further Studies of Natural Reforestation in the Donner Ridge Burn," Proceedings, *Annual Tall Timbers Fire Ecology Conference* (1974): 195–200; C. E. Bock, M. Raphael, and J. H. Bock, "Changing Avian Community Structure during Early Post-Fire Succession in the Sierra Nevada," *Wilson Bulletin* 90 (1978): 119–23; J. H. Bock, C. E. Bock, and M. Raphael, "A Comparison of Planting and Natural Succession after a Forest Fire in the Northern Sierra Nevada," *Journal of Applied Ecology* 15 (1978): 597–602.
5. Richard D. Taber, "Deer Nutrition and Population Dynamics in the North Coast Range of California," *Transactions of the North American Wildlife Conference* 21 (1956), 159–72.

6. Robert McCabe, *Aldo Starker Leopold: 1913–1983* (Washington, DC: National Academy of Sciences, 1990), 237–55.
7. Luna B. Leopold, "Ethos, Equity, and the Water Resource," *Environment* 32, no. 2 (1990): 16–41.
8. N. L. Bradley, "Sand County Restoration."
9. Darling, *Pelican in the Wilderness*, 66.
10. A. Carl Leopold, R. Andrus, A. Finkeldey, and D. Knowles, "Attempting Restoration in Wet Tropical Forests in Costa Rica," *Forest Ecology and Management* 142 (2001): 243–49.
11. Luther J. Carter, "The Leopolds: A Family of Naturalists," *Science* 207 (1980): 1051–55.

Chapter Ten

1. Wendell Berry, *The Art of the Commonplace* (Washington, DC: Counterpoint, 2002), 45.
2. Berry, *Art of the Commonplace*, 44.
3. Loren Eiseley, *The Unexpected Universe* (San Diego: Harcourt, Brace & World, 1969).
4. Fassett, *Spring Flora of Wisconsin*.
5. Carter, "The Leopolds."
6. A. Leopold, *Sand County Almanac*, viii.
7. Wes Jackson, *From Swords into Plowshares*, August 1981.
8. Robert Michael Pyle, *Sky Time in Gray's River* (Boston: Houghton Mifflin, 2007), 99.
9. A. Leopold, *Sand County Almanac*, "Marshland Elegy," 95.

Appendices

* Robert McCabe, *Aldo Leopold, the Professor* (Amherst, WI: Palmer Publications, Inc., 1987).
1. Steven R. Brower, "Research Report: The Starker-Leopold Family," 1980.
2. Brower, "Research Report."
3. Gilberto Espinosa and Tibo J. Chavez, *El Rio Abajo* (Portales, NM: Bishop, n.d.), 57.
4. Espinosa and Chavez, *El Rio Abajo*.
5. Darling, *Pelican in the Wilderness*, 85.

Index

Note: Page numbers in *italics* indicate images and captions.

Adams County, 143
airplanes, 45–46
Aldo Leopold Center, 189
Aldo Leopold Foundation
   and the Bradley Study Center, 258
   and land restoration efforts, 227–28, 229, 239, 244
   and the Leopold Fellows Program, 222, 224
   and the moist prairie, 206
   origins of, 215, 227–28
   and phenology observations, 238, *238*
Aldo Leopold Memorial Reserve (LMR), *187*, 216–17, 220, 222, *231*, 232
Aldo Leopold Oak, 214
Aldo Leopold Shack Foundation, 220, 222, 228. *See also* Aldo Leopold Foundation
Alexander, Emma, 6–8
Alexander, Jacob, 6–8, 195
alfalfa fields, 229, 232
aluminum pots, 24
American Wildlife Institute, 286
Anchor, Carl, 217
Anchor, Eleanor, 217
Anchor's Island, 66, 152
Anchor's Woods, 25, 66, 91, 129, 147, 152, 208
Anderson, Frosty, 286

anemone, 201
annual oats (*Avena fatua*), 219
apple trees, 3–4, 6, 10, *11*, 97, 130, *132*, 203, 213
archery
   and hunting, 140–44, 144–47, 152, *162*, 169–75
   and neighbors of the Shack, 8
   and origins of the Shack, xi
   and roving, 160–63
Archibald, George, 234–35
arrowheads, 125–26
arrow-making, 159–60
Art of the Commonplace (Berry), 268
aspen (*Populus tremuloides*), 183, 193
Attig, John, 177
Austin, Mary, 177

Badger Ordnance Works, 222
Badlands National Park, 257
Bakhtin, Victor, 229–30, *230*
banding birds, 59–60, 247
Baraboo, Wisconsin, 19, 32, 61, 74, 82
Baraboo Bakery, 32
Baraboo Hills, 178, 198, 230
Baraboo River, 169
barn (original Shack), *4*, 8
barred owls, 92–93
Barrows Bluff, 9, 177, 180–81

Baxter, Caroline, 5–6
Baxter, William, 5–6, 16, *17*, 203, 206
Baxter family, 130
beach, 7, *69*, 117–20, *121*
Bee Tree, 7, 51–54, *53*, *54*, 127
Belén, New Mexico, 294
Bellrose, Frankie, 94
Bent Road Prairie, 232
Bergere, Alfred, 295–96
Berry, Wendell, 268
Birch Row, 7, *29*, 30–31, 200
birch trees (*Betula papyrifera*), 26, 30–31, 63, 104, 183, 191–92, 200, 208
birds
    banding birds, 59–60, 247
    and courtship flights, 87–88
    as family pets, 78, 80–85, *84*, *151*, 280–84, 284–86, 286–89
    and field trips, 97
    gardening for, *131*
    and the Good Oak, 71
    and Lake Chapman, 124
    and land restoration efforts, 247
    and Meat Rock, 91–94
    and migrating species, 89–91
    monitoring programs, 242
    and rain forest project, 261
    and return of rare species, 234–35, 254–55, 277
    and road kill, 100
    and "Shack West," 265–66
    and student field trips, 96–98
    and summertime, 103
    and winter, 49–50, *52*, 54–57, *56*
    See also specific species
birdsfoot violets (*Viola pedata*), 103, 191
Birmingham and Hixon, 14, 40, 44
black ash (*Fraxinus nigra*), 183

Blackhawk (Sauk chief), 9
black oaks (*Quercus velutina*), 188, 214, 228, 233, 242
bloodroot (*Sanguinaria canadensis*), 87
bluegrass, 195
blue lilac bush, 190
Bock, Carl, 250–51, 252
Bock, Jane, 250–51, 252
bogs, 182. *See also* marshes
boneset (*Eupatorium perfoliatum*), 133
botany, 182, 273
bottomlands, 145–46
bow-making, 157–59, *159*
Bradley, Charles Crane
    and airplanes at the Shack, 45–46
    and the Aldo Leopold Foundation, 215
    and the Bradley Study Center, 217–21, *255*
    Charlie's Wood and Prairie, 228
    and class field trips, 96
    and land restoration efforts, 230, 246
    and the Leopold Fellows Program, 222
    and music at the Shack, 34
    and "Shack West," 265
    sketch of oak tree, *218*
Bradley, Nina
    and airplanes at the Shack, 45–46
    and the Aldo Leopold Foundation, 215, 228
    and the Aldo Leopold Oak, 214
    and archery, 161, 163–64, 169, 174
    and the beach, 117
    and bird watching, 89, 124, 234–36

and the Bradley Study Center, 217, 219–21, 234–36, *255*, 255–58
and Carl's falconry, 137
and Charlie's Wood and Prairie, 228–29
death, 257–58
and family dynamics, xi, xii
and family pets, 281, 285, 286–89
and father's death, 212
and fireplace reconstruction, *39*, 40
and firewood gathering, 53, *57*, 58
and Gus, 152
and land restoration efforts, 224–27, 230, 232–33, 244, 246
and Leopold benches, 110–11, 115–16
and the Leopold Fellows Program, 221–24
marriage, 127
and music at the Shack, *33*, 34
and owls, 93
and phenology observations, 237–38, *238*
and prairie restoration efforts, 198
and reconstruction of the Shack, *4*, *15*
and road kill, 99–100
scholarly career, xiv, 273
and "Shack West," 265–66
and the Slide Hill Bench, *116*
and summertime at the Shack, *105*, 109
and tree planting, 77, 79–80
and vandalism of the Shack, 61
and winter at the Shack, 49–50, 67–68
Bradley Study Center
and Charlie's Wood and Prairie, 228–30
garden at, 221
and history of Shack area, 186–87
and the ideology of the "Shack," 255–58
and land restoration efforts, 217–21, 232
and the Leopold Fellows Program, 224
and phenology observations, 237, *238*
and sandhill cranes, 234–35
white oaks at, *187*
in winter, *256*
woodpile at, *255*
bridges, 122, *123*, *185*
brown-eyed Susans (*Rudbeckia hirta*), 191, 197, 257
brush removal, 244
bunkhouse, 13, *14*
bunks, 16, 109–10
burr oaks (*Quercus macrocarpa*), 186, 188, 213–14, 228
butterfly weed (*Asclepias tuberosa*), 191, *192*

Camp Randall, 166
Canadian geese, 94–95
carbon-14 dating, 126, 182
Cardamine, 209–13
Case, Roy, 144, 173
cedar trees
and arrow making, 159
and fireplace reconstruction, 40, 43–44, 61
and firewood gathering, 58, *58*

cedar trees (*continued*)
    and first visit to Shack property, 5, *6*
    and reconstruction of the Shack, 19
    and vegetable garden at Shack, 203, 205
cement, 38, *39*, 42–43, 44, 48
Charlie Bradley's woods, 228–30
Chavez, Tibo J., 295
Chicago Daily News, 166
Chicago Tribune, 164
chickadees, 54, 56, *56*, 59, 62, 247, 265
chimneys, *4*, *14*, 40, *42*, *47*, 48, 265, 281
cinquefoil (*Potentilla arguta*), 206
Civil War, 25
Clandeboye Bay, 235, 286
clay flooring, 17–19, 42
Clay Hill, 5, 7, 69, 75, 126, 177, 180–81
cliffrake (*Pellaea*) rock fern, 92
climate change, 177–78, 180, 183–84, *238*
coffee pots, 24
Coleman, Catherine, 12
Coleman, Reed, 31, 110, 216–17, 224
Coleman, Thomas, 12, 74, 79, 110, 203, 216
Coleman Prairie, 232
Collins, John, 227
compass plant (*Silphium laciniatum*), 102, *102*, 198, 256–57
coneflowers (*Echinacea* sp.), *192*, *221*
conifer forests, 182–83, 242. *See also* pine trees; *specific species*
Conservation Commission, 210
Conservation District, 73
conservation easements, 266, 275
controlled burning, 187, 201, 206, 225–26, 230, 232–33, 243, 251
cooking
    and fireplace reconstruction, 38, *43*
    and game birds, 150
    and reconstruction of the Shack, 22–24
    sketch of camp cooking, *23*
    and sourdough pancakes, 103–4
    and summertime at the Shack, 107–9
Cooper's hawk, 134–36, *135*
corn fields
    and class field trips, 96
    and first visit to Shack property, *3*, 5–6
    and geese, 94–95
    and geology of the Shack area, 180
    and land restoration efforts, 195–201, 246, 247
    and restoration of native plants, 195–201
    and the Shack garden, 132
    and summertime at the Shack, 101
    and tree planting, 27, *27*, *29*, 30, 75, 80
Cortez, Hernando, 293
cottonwoods (*Populus*), 113, 118, 183, 269, 275–76
"Cottonwood Tree" (Leopold), 275–76, *276*
County Truck T Prairie, 232
cowboy ballads, 34
cows, 10–11
cow's horn, *86*, 106

crows, 150, 279, 280–84
cup plant (*Silphium perfoliatum*), 198
Curtis, John, 189, 201, 215, 217, 224

daisies (*Erigeron* sp.), 206
D and RG Railroad, 294
Darling, Fraser, 74, 214, 257, 296
Davis, Jefferson, 25
deer and deer hunting, 95–98, 144–47, 152–53, 170–75, 251–52
Delta Waterfowl Research Station, 286
dense blazing star (*Liatris spicata*), 191
Devil's Lake, 230
Driftless Area, 180, 242–44, 267
drought, 31, 102, *233*, 247
dune plant (*Hudsonia tomentosa*), 207
Dunwiddie, Peter, 203
Dutch elm disease, 30
Dutch oven, 22–23, *23*, 40–41, *43*, 60, 103–4, 109
Dyksterhuis, E. J., 198–99

early forests, 182–83
The Earth Knows My Name (Klindienst), 190
EBL Prairie, 220, *221*, 232, 246, 247, 256
Ebner, Adelbert, 74
"ecological landmarks," 25
Ecological Monographs, 198–99
Eiseley, Loren, 271
Elder, Bill, 46, 127, 286–87
Elder, Nina. *See* Bradley, Nina
Eleocharis, 118
Ellarson, Bob, 217

elm trees
  and the beach, 118
  and family pets, 85–86
  and firewood gathering, 58–59
  and first visit to Shack property, 2–3, 5
  and land restoration efforts, 197
  and Leopold benches, 111, *112*, 115–17, *116*
  and neighbors of the Shack, 10
  and outdoor shower, *194*, 194–95
  and the tree house, 110
  and tree planting, 26–27, *28*, 30
  and vegetable garden, *204*
  and vegetation phases, 183
El Rio Abajo (Espinosa), 295
Elton, Charles, 74, 213–14
emigrant road, 5, *7*, *50*, *51*
Emlen, Johnnie, 90, 164, *165*
Emlen, Virginia, *165*
Environmental Protection Agency (EPA), 226
Espinosa, Gilberto, 295
European agricultural practices, 188

falconry, 133–40, *135*
Fallas, Memo, 260
fall season, 130–33, 144–47, 152
Farm Hour radio show, 197
Fassett, Norman, 189, *233*
Feeney, Bill, 134, 137
fences and fencerows, 197, 204, *204*, 264–65
field trips, 96, *96*
fireplaces, 13, 37–44, 46–48, *47*
fires, 187, 199–201, *200*, 209–13. *See also* controlled burning
firewood, 51–59, *53–58*, 69–72
Fish and Wildlife Service, 94
fishing, 1, 12, *122*, 122–24

Flader, Susan, 235
Flicky (German shorthaired pointer), 155, 209, 211–12, 225
Flicky (springer spaniel), *21*, *22*, *28*, 51, 74, *123*, 147, *165*
flooding
  and the beach, *120*
  and geology of the Shack area, 177–78
  and hunting, 141–42
  and reconstruction of the Shack, 21–22, *22*
  and Slough, 62–63
  and springtime at the Shack, 78–80
  and tree planting, 26
floodplain, 7, 9, 22, *120*
Flowering spurge (*Euphorbia corolata*), 197
Fluminea (Manitoba crow), 287–89
Fort Winnebago, 5, 25
fossilized pollen, xiv, 182, 184, *185*, 187, 273
foundation of the Shack, 42–43
Fox Indians, 5
Franklin, Benjamin, 37–38
Frank's Prairie, 232
Freeman, Scott, 227, 266
Freeman, Susan Leopold, 266
Fremont Lake, 252
Friendship Foundation, 226
fringed gentian (*Gentiana crinata*), 133
Funkia plants, 190
Future Leaders Program, 244

gardens
  for bird forage, *131*
  at Bradley Study Center, 221
  and fall at the Shack, 130–32
  at Luna Mansion, 294
  and neighbors of the Shack, 9, 10, *11*
  and reconstruction of the Shack, 13
  vegetable garden, 9, 130–32, *131*, *132*, 203–5
geology of Shack area
  and class field trips, 97
  and early forests, 182
  and fossilized pollen, 273
  and glacial features, 180–81
  key features of Shack area, 177–78, *178*, *179*
  and the Leopold Fellows Program, 221
  and maple trees, 203
  and music at the Shack, 34
  and the sand blow, 124–27
  See also glacial formations
Georgia O'Keeffe Foundation, 296
Gilbert farm, 9–10, 12, 143, 177, 180, 210, 214, 216
Gilbert's Island, 7
glacial formations
  and the Driftless Area, 242–44
  and family pets, 287
  and geology of the Shack area, 177–78, *178*, *179*, 180–81
  and Meat Rock, 89, 92
  and the sand blow, 124–26
  and vegetation phases, 183–84
"goat prairies," 233
Good Oak, *51*, 69–72
"The Good Oak" (essay), 69, 71–72
Gordon, Eddie, 129
graduate students, 95–98, *96*, 98–100, 250–52, 286
grapes, 31, 133, 173, 221

gray dogwoods (*Cornus racemosa*), 206
Gray's Manual of Botany, 26
Great Depression, 2, 13
great horned owl, 93, 137–40, *138*
Great Marsh, 7, 235
Great Plains, 187, 256–57
Green Bay ice lobe, *178*, *179*, 180–81
Green River, 252
grouse hunting, 143–44
Gunderson, Mr., 147
Gus (German shorthaired pointer), *148*, *149*
 and author's pets, 80, 82–83, 85, *151*
 and hunting, *142*, 143, 147–55
 sketch of, *154*
 and summertime at the Shack, 103, 105
 and tree planting, 74
 and winter at the Shack, 68, *69*
Gus's Island, 65
Guthrie, A. B., 83–85

Haley, Billy, 78
Hamerstrom, Fran, 97, 98–100, 207
hard maples (*Acer sacharum*), 202–3, 204
harrier hawks, 100
Hawkins, Art, 148
hawks, 133–40
hay bedding, 16, 41–42
haystacks, *17*
hemp, 10, *131*
Henika, Franklin, 147, 173
hepatica, 201
Highway 12 Prairie, 232
Hochbaum, Hans Albert, *54*, 97, 286

honey, 8, 52, 104
Horsemeyer, Mr., 161
Huffaker, Buddy, 229, 239, 244, 258
hunting
 and Aldo Leopold, 292
 and archery, 140–44, 144–47, 152, *162*, 169–75
 deer, 144–47
 and Delta Waterfowl Research Station, 287–88
 and falconry, 133–40
 and family dynamics, xi
 and first visit to Shack property, 5
 and Gus, 147–55
 and origin of Shack property, 1
 pheasants, *142*
ice floes, 65–66
Iltis, Hugh, 258
immigrants, 70
Important Bird Area Program, 239, 242
inaugural visit to property, 2–8
Indian grass (*Sorghastrum nutans*), 197
International Crane Foundation, 223, 226, 234–35
islands, 63–64, *64*, 106–7, 232

jack pines (*Pinus banksiana*), 183, 206–7
Jackson, Wes, 276
Jacob's ladder, 201
Jagdschloss (hunting lodge), 8
joe-pye weed (*Eupatorium maculatum*), 133
Johnstown Moraine, *178*, 180–81, 186, 191–92
Jones, Elizabeth, 190
journal, 46, 72, 95, 212–13
Joy Camp, 281–83

Kansas gayfeather (*Liatris pycnostachya*), 197
Kenney, Bergere, *161*
kettles, 22, *23*, 38, 41, *43*
Klindienst, Patricia, 190
Knight, Rick, 222
Konrad, Liegel, 222–23
Koshollek, Alanna, 239, 242–43
Krutzman, Mr., 147
Kumlien Club, 97–98

La Fleche, Andre, 287, 289
Lake Agassiz, 287
Lake Chapman, 7, *185*
   and the Aldo Leopold Memorial Reserve, 216
   bridge supports, *123*, *185*
   and land restoration efforts, 183–84
   and neighbors of the Shack, 11, 110
   and the Regan fire, 210–11
   and tree planting, 74–76
Lake Fremont, 266
Lake Mendota, 203, 237
Lake Winnipeg, 235, 286
land conservation, xi
Land Institute of Salinas, Kansas, 102, 226
Laubach, Steve, 5, 217
lead plant (*Amorpha canescens*), 191, *192*
lecture series, *223*
Leopold, A. Carl, *262*
   and airplanes at the Shack, 45
   and the Aldo Leopold Memorial Reserve, 217
   and the Aldo Leopold Oak, 214
   and archery, 157, 161, *161*, 163, 169, 173–74
   beach photo, *120*
   and birds at the Bradley Study Center, 235
   and bird watching, 89, 124
   and connection with the land, 275
   Costa Rica rain forest project, 258–63, *259*
   and falconry, 133–40, *135*, *138*
   and family background, 270
   and family dynamics, xi
   and family pets, 285
   and fireplace reconstruction, 48
   and firewood gathering, 53, *54–57*, 56–57
   and first visit to Shack property, 6
   and fishing, *121*
   and Gus, 151–52
   and hunting, 142–43
   and the ideology of the "Shack," 266
   at Lake Chapman, *123*
   and land restoration efforts, 189, 208–9, 233, 244
   and Leopold benches, 117
   and maple trees, 202
   and music at the Shack, 33–34, *35*
   and neighbors of the Shack, 10
   and outdoor shower, 194
   and owls, 92–93
   and phenology observations, 237
   and reconstruction of the Shack, 12–13, 18, 21, 24
   and road kill, 99–100
   scholarly career, xiv, 273
   and "Shack West," 265
   and summertime at the Shack, *108*, 109
   and trail building, 127
   and the tree house, 110
   and tree planting, 31, 77

and vegetable garden, 204
and winter at the Shack, 67–68, *69*
Leopold, Aldo
   and airplanes at the Shack, 45
   and the Aldo Leopold Foundation, 215
   and the Aldo Leopold Memorial Reserve, 216
   ancestry, 293
   and archery, 156, *157*, 157–60, 161–63, 163–65, *165*, 169–71, 173–75
   and arrowhead find, 125
   and the beach, 117–19
   and bird banding, 60
   and bird watching, 90–91
   and Carl's falconry, 133–37, 139–40
   and class field trips, 95–98, *96*
   and connection with the land, 275–77
   and cutting of the Good Oak, 69–72
   death, 211–12
   and family background, 268–69, *270*, 296–97
   and family dynamics, xi–xiv
   and family pets, 77–80, 82–83, 85, 87, 279, 281, 284, 285, 286, 289
   and fireplace reconstruction, 38–44, *42*, 47, *47*, 48, 49
   and firewood gathering, 52–53, *53*, 55, *57*, 58
   and geese, 94–95
   on glacial features, 92
   and Gus, 147–55, *149*
   and history of Shack area, 186
   and hunting, *xiii*, 141–44, 144–47
   and the ideology of the "Shack," 267
   and land restoration efforts, 176, 188–90, 195–99, 207, 208–9, 224, 245–47, 249
   and Leopold benches, 111–13, *114*, 115–17
   and maple trees, 202
   memorial oak for, 213–14
   and the moist prairie, 206
   and music at the Shack, *32*–34
   and native plantings in Shack yard, 191–93, *192*
   and neighbors of the Shack, 8–10
   and phenology observations, *237*, *238*
   and reconstruction of the Shack, 1–2, 12–14, *13*, *14*, *15*, 17–19, *21*, *22*, 22–24
   and the Regan fire, 209–13
   and sandhill cranes, 234–35
   and scholarly careers of children, 272–73
   and the Slide Hill Bench, *116*
   with students, *196*
   and summertime at the Shack, 101–2, 103–9, *106*, *108*, 110
   and trail building, 127–28
   and tree planting, 24–31, *28*, *29*, 73–76, *75*
   and vandalism of the Shack, 61
   and vegetable garden, 203–5
   and winter at the Shack, *52*, *63*, 66–67
Leopold, Betty, 251
Leopold, Carl (grandfather), 157, 291–93
Leopold, Carl (uncle), 117, 151, 174, 291–92

Leopold, Caryl, 118, 124
Leopold, Charles J. J., 291
Leopold, Dolores Bergere, 117–18, 151
Leopold, Estella B. (author)
    and airplanes at the Shack, 45
    and archery, 161
    and bird forage garden, *131*
    charcoal sketches, *23*, *154*
    and family pets, 77–87, *81*, 281–83
    and fireplace reconstruction, *39*
    and firewood gathering, *57*
    and fishing, *64*
    and hunting, 142, *142*, 146
    and Leopold benches, *112*
    and music at the Shack, *35*
    name change, 72
    and reconstruction of the Shack, *4*, *15*, *21*
    and road kill, 99–100
    scholarly career, xiv
    and the Slide Hill Bench, *116*
    and summertime at the Shack, *105*
    and vegetable garden at Shack, 131–32
    and winter at the Shack, *50*, *69*
Leopold, Estella Bergere (author's mother), *278*
    and the Aldo Leopold Memorial Reserve, 216–17
    ancestry, 293–97
    and apple orchard, 130–33
    and archery, 156, 158, 160–61, 163–65, *165*, 166–68, *167*, *168*, 169–74
    and arrowhead find, 125
    and the beach, 117, 119
    and bird watching, 91, 235–36
    and the Bradley Study Center, 220–21, 235–36
    and Carl's falconry, 136–37
    and class field trips, 97
    and cutting of the Good Oak, 70–72
    at EBL Prairie, *221*
    and family background, 269, *270*, 271–72
    and family dynamics, xii–xiv
    and family pets, 80, 83, *84*, 86–87, 279, 281, 283–84, 289
    and fireplace reconstruction, 40–41, 44, 46–47
    and firewood gathering, 53, 55–56, *57*
    and first trip to shack property, 3
    and fishing, *12*
    and geese, 94–95
    and Gus, 147–48, 150, 152, 155
    and hunting, *xiii*, 141, 143–44, 146–47
    and the ideology of the "Shack," 267
    and land restoration efforts, 190, 246, 247
    and Leopold benches, 111–15, *114*
    and memorial oak for Aldo, 213–14
    and music at the Shack, 32–34, *36*, 37
    and native plantings in Shack yard, 190–92, *192*
    and neighbors of the Shack, 8, 10
    and outdoor shower, 195
    and owls, 93
    and reconstruction of the Shack, 14, 17–18, 19, *21*, 21–24
    and the Regan fire, 209–12
    and the Sage Hen Field Station, 251

and summertime at the Shack,
103–9, *105*, *107*
and trail building, 127
and tree planting, 27–29, *29*, 31,
73–74, 77
and vandalism of the Shack, 60
and vegetable garden, 203
and vegetable garden at Shack,
132
and winter at the Shack, 49, *51*,
63, 67
and wood cutting, 77
Leopold, Fritz, 251
Leopold, Luna
and archery, *157*, 161, *161*,
163–64, 169, 172
and family background, 272
and family dynamics, xi
on father's kindness, 293
and fireplace reconstruction,
37–38, 40, *41*
and firewood gathering, 53
and fishing, *122*
and Gus, 152
and hunting, 141, 143, 147
and the ideology of the
"Shack," 266
and land restoration efforts, 244
and Leopold benches, 111
and memorial oak for Aldo,
213–14
and music at the Shack, *33*–37
and reconstruction of the Shack,
4, 13, 16, 18–19
scholarly career, xiv
and "Shack West," 263–64, 266
and summertime at the Shack,
105, 109
and trail building, 127
and the tree house, 110–11
and tree planting, *28*
in Wyoming, 252–55, *253*
Leopold, Lynn Bradley, *187*,
235–36, 260
Leopold, Madelyn, 266
Leopold, Starker, *250*
and the Aldo Leopold
Foundation, 227–28
and the Aldo Leopold Memorial
Reserve, 217
and the Aldo Leopold Oak, 214
and archery, 158, 161–63, *162*,
163–65, *165*, 169–70, 174–75
and family background,
290–91
and family dynamics, xi
and fireplace reconstruction,
38, 40
and firewood gathering, 53
and Gus, 152
and hunting, 141–43, 145–47
and land restoration efforts, 190,
197, *200*, 201, 244, 246
and music at the Shack, *33*, *35*
and reconstruction of the Shack,
13–14, *14*, 16, *18*, *21*
and the Sage Hen Field Station,
249–52
scholarly career, xiv, 272–73
and trail building, 127
and tree planting, *26*, 74
Leopold, Susan, 227
Leopold benches, *20*, *28*, *105*,
111–17, *114*
Leopold Center, 177, 191, *192*,
238, *243*, 244
Leopold Desk Company, 117, 157,
291–92
Leopold Fellows Program, 221–24,
222, 227

Leopold Foundation. *See* Aldo Leopold Foundation
Lewis, Joe, 46, 49, 80, 95
Liatris spicata, 191
Liegel, Konrad, 222
livestock, 10–11
Living the Land Ethic (Laubach), 217
Long Marsh, 232
Lord, Marie Leopold, 27, 74, *75*
Los Arboles, 260–62
Los Lunas, 294
lousewort (*Pedicularis*), 206
lowlands, 26. *See also* bottomlands; marshes
L. R. Head Foundation, 216–17, 218
lumber, 66–67
Luna, Antonio Jose, 294–95
Luna, Eloisa, 294
Luna, Isabel de Baca, 294
Luna, Jose Enrique, 294
Luna, Salomon, 295
Luna, Tranquilino, 294
Luna Mansion, 294
Luthin, Charlie, 222, 228
Lysimachia quadrifolia, 271

Madison, Wisconsin, 8
Mahler, Louis, 126
Main, Dorothy Turner, 28
Main, John S., 28
"Man's Leisure Time" (Leopold), 156
mantelpiece, 40
Maple Bluff, 202
maple trees
    and the beach, 119
    and bird watching, 89
    and "bridge" to island, 65
    and furniture making, 292
    hard maples (*Acer sacharum*), 202–3, 204
    and hunting, 146
    and Leopold benches, 111
    soft maples (*Acer saccharinum*), 202–3, 208, 225
    and summertime at the Shack, 104, *105*
    and tree planting, 26
marshes
    and the Aldo Leopold Memorial Reserve, 216
    and birds at the Bradley Study Center, 235
    and early forests, 182
    and family pets, 286–87
    and first visit to Shack property, 5
    and geese, 94–95
    and history of Shack area, 184
    and hunting, 174
    of Lake Chapman, 121
    and land restoration efforts, *231*
    and neighbors of the Shack, 9
    and the Regan fire, 210–11
    and springtime flooding, 79
    and tree planting, *29*, *30*
"Marshland Elegy" (Leopold), 235
marsh marigold (*Caltha palustris*), 90–91
mattresses, 41–42
Mayer, Teresa, 222, 237–38
McAleese, Kevin, 230
McAndrews, Jock, 187–88
McPhee, John, 100
Mead, Howard, 216
Meat Rock, 91–94, 115, 180
Meine, Curt, *223*
merlin (pigeon hawk), 137–40
Mertensia, 191, *192*
migrations, 94, 237, 277

milk cows, 10–11
milkwort (*Polygala*), 206
mineral content of soil, 226.
    See also soil fertility
moist prairie, 205–6
moraines, 69, 124–26, *178*, *179*, 180–81, 186
mosquitoes, 19–21
Mossman, Mike, 222, 239
mountain ash, 27
Muir, John, 70, 185–86, 218
Muir family, 70
mulberry tree (*Morus nigra*), 203–4
Murrish, Patty, 212
music
    and class field trips, 97, 223, *223*
    and early years at the Shack, 32–37
    and family background, 272
    and family pets, 83
    and Gus, 150
    and hunting, 174
    and Sage Hen Field Station, 251
    and summertime at the Shack, 108–9
    and the tree house, 111
muskrats, 62–63
mustard plants (*Draba* sp.), 195–96
My Healthy Woods (Swenson and Koshollek), 242–43

Napoleon bench, 119
National Archery Association, 164
Native Americans, 5, 9
native plants
    and the Bradley Study Center, 219
    and first visit to Shack property, 8
    flowering natives, 105–6, 191, *192*
and land restoration efforts, 176, 188–92, 195–201, 226–27, 229, 242
    in old cornfield, 195–201
    and tree planting, 24–32
    See also specific species
Natural History, 234
N-Bar Ranch, 37
neighbors, 8–12
Nelson, Barbara, 254
Nelson, Carrie, 227
New England aster (*A. novae-angliae*), 133
New Mexico, 294
Northwest Furniture Company, 291–92
Norway pines, 26–27, 31
nurseries, 226

oak trees
    and brush removal, 244
    and land restoration efforts, 226, 233–34
    "oak openings," 186, 188
    and vegetation phases, 183–84
    See also specific species
oak wilt, 233
Ochsner, Ed, 2, 8, 104, 147
Ohio spiderwort (*Tradescantia ohiensis*), 103
onions, 132
orchards, 3, 5, 10–11, 40, *58*, 130–33, 203–5, 294
orchids, 30
original Shack, 3–8
ornithology, 99
Otero, Adelida, 295
Otero, Antonio Jose, 294–95
Otero, Eduardo, 295
Otero, Eloisa Luna, 295

Otero, Manuel B., 295
Otero, Nina, 295
Otter Pool, 7
otters, 236
outhouse. *See* The Parthenon (privy)
owls, 91–94

Packard Foundation, 228
palynology, 273
panic grass (*Panicum* sp.), 119, 196
The Parthenon (privy), 7, 16–17, *18*, 30, 103, *121*, 197, 205
pasqueflower (*Anemone patens*), 74, 87
Paulson, Reuben, 285
Pedro (crow), 78, 80–85, *84*, *151*, 284–86
peppergrass (*Lepidium*), 195
pheasants, 141–42, *142*
phenology observations, 90–91, 105, 190, 212–13, 222, 237–38, *238*
phoebes, 90–91
photography, 31
pigeon hawk (merlin), 137–40
pileated woodpeckers, 91
pine-birch period, 183
Pine Island, 25
pine trees, 24–32, 67, *106*, 208–9, *240*, *241*, 242, 247. *See also specific species*
Plummer's Slough, 9, 143
Poco (squirrel), 79–80, *81*, 85–87, *86*
pollen, 125, 182–84, 185, *185*, 187, 273
pollen, fossilized, xiv, 182, 184, *185*, 187, 273
polygala (milkwort), 30
Polytrichum (haircap), 30
ponderosa pines, 263–64, *264*

Portage, Wisconsin, 5, 8–9, 25
potatoes, 130–31
"Prairie Birthday" (Leopold), 198
Prairie Birthday site, *102*, 233
prairie chickens, 99–100
prairie cinquefoil (*Potentilla arguta*), 197
prairie clover (*Petalostemon purpurea*), 197
prairie fires, 203
prairie flora
 and the Bradley Study Center, 257
 and fire lanes, 201
 and land restoration efforts, 189, 195–201, 219–20, 224–27, 229–30, 230–33, *231*
 and summertime at the Shack, 102
 and vegetation phases, 184
 See also specific species
prairie phlox (*Phlox pilosa*), 197
prescribed burns, *202*, 217, 225, 243–44. *See also* controlled burning
prickly-ash (*Zanthoxylum americanum*), 127–28
privy. *See* The Parthenon (privy)
Proceedings of the National Academy of Sciences, 237
prothonotary warblers, 124
puccoon (*Lithospermum caroliniense*), 191
purple coneflower (*Echinacea pallida*), *192*, 197
Pyle, Robert, 100, 277

quail, 141, 196–97
Quintanilla, Norberto "Beto," 175

rabbits, 70
Rahr, Guido, 148, 155

railroads, 101–2, 188, 294
rain forests, 258–63
Rand, C. W., 291
Rand-Leopold Desk Company, 291
Randolph, Jean, *39*
reconstruction of the Shack, 1–2, 3–9, *4*, 12–24
recycled and repurposed materials, 13–14, 16, 44. *See also* river flotsam
red cedars, 5, 203, 205
red oaks (*Quercus rubra*), 228, 242
red pines (Norway pines), 26–27, 31, 73–74
Regan, Mr., 210–11
Regan Marsh, 210
restoration efforts
    and the Aldo Leopold Foundation, 227–28
    and the Aldo Leopold Memorial Reserve, 216–17
    and avian species, 234–36
    and the Bradley Study Center, 217–21
    and Charlie Bradley's woods, 228–30
    and the Driftless Area, 242–44
    and the ideology of the "Shack," 246
    and the Leopold Fellows Program, 221–24
    and maple trees, 201–3
    and the moist prairie, 205–6
    and oak forests, 233–34
    and the original cornfield, 195–201
    other projects on Leopold lands, 239–42
    and phenology, 237–38
    and pine plantings, 208–9
    and prairie building, 224–27
    and restoration ecology field, 188–90
    restored vegetation areas, 230–33
    and the sand blow, 206–7
    and the Shack yard, 190–95
    and tamaracks, 207–8
    and the vegetable garden, 203–5
Restoration Notes, 225
ridges, 177–81, *179*
Riggs, Maynie, 46
Riley Game Cooperative, 141, *142*
Ring Around, 7, 30
river birch trees, 30–31, 63
river flotsam, 13–14, 66–67, 119
River (Levee) Road, 174, 205–6
road kill, 98–100
Roark, Ray, 27–28
Rogers, Barbara, 212
Rogers, Keena, 271
rosinweeds (*Silphium*), 198
Ross, Anne, 228
Round River (Leopold), 155, 176
Roundup (glyphosphate), 225
roving, xi, *157*, 160–63, *161*, 169
Runge, Thusnelda, 291

Sage Hen Field Station, 249–52
Sammy (crow), 280–84
sand blow, *6*, 124–27, 206–7
sand burrs (*Cenchrus* sp.), 195
Sand County, 215, 224
A Sand County Almanac (Leopold), xi, 2, 69, 72, 93–94
Sand County Foundation, 217–18, 220, 222, 230
Sand Hill, *6*, 7
    and bird watching, 89

Sand Hill (*continued*)
   and class field trips, 96
   and first visit to Shack property, 5
   and geology of the Shack area, 177, 180–81
   and pines, 208–9
   and trail building, 129
   and tree planting, 79
   and vegetable garden, 203, 205
   and winter at the Shack, *51*
sandhill cranes, 234–35, 254–55, 277, 286
sand-point pump, 19, *20*
Santa Fe, New Mexico, 34–37
Sauk County, 178, *178*, 178–80, 215, 222, 230–33, *231*
Sauk Indians, 5, 9
Sauk Prairie, 233
Sauk Prairie Remembered: A Vision for the Future (Bakhtin), 229–30, *230*
savannas
   and bird watching, 88
   and first visit to Shack property, 5
   impact of agricultural practices, 188
   and oak plantings, 213, 218
   and oak species, *187*
   and restored areas in Sauk Country, 230
   and tree thinning programs, *241*, 242
   and vegetation phases, 184–85
sawmills, 25
screech owls, 93, 98
seedlings. *See* tree planting
serviceberry (*Amelanchier arborea*), 193
"Settlement House," 8–9
Severinghaus, Ed, 285

"the Shack" ideology, 245–67
Shack Prairie, 7, 232
"Shack West," 263–67, 274–75
shagbark hickory (*Carya ovata*), 228
sheep ranching, 294–95
shepherd's purse (*Capsella*), 195
shingles, 44
shooting stars (*Dodecatheon media*), 206
shotguns, *xiii*, 141–43, 152–53
shower, outdoor, *194*, 194–95
shrubs, 27
skating, 62
skiing, 49–50, *50*, *51*, 68, *69*, 70, 129
skunk cabbage (*Symplocarpus foetidus*), 90–91
sleds, 51–53, 55, *55*, 57, *57*, *58*
Slide Hill bench, 115–17, *116*, *128*
the Slough
   and the beach, 117
   and bird watching, 89
   and deer hunting, 146
   and firewood gathering, 53–54, *53–54*, 57, *57*
   and first visit to Shack property, 3
   and land restoration efforts, 232, 246
   and Leopold benches, 113, 115, *116*
   and maple trees, 202
   and neighbors of the Shack, 9
   and owls, 91–92
   and phenology observations, 213
   and trail building, 127
   and winter at the Shack, 51, *52*, 62–64, 66–67
Smith, Caryl Leopold, 266
soft maples (*Acer saccharinum*), 202–3, 208, 225
Soil Conservation Service, 199

soil fertility, 188, 226, 246, 251–52
sorghum, 10
sorrel (*Rumex acetosella*), 195
sourdough, 103–4, 109
Sowls, Grace, 286
Sowls, Lyle, 286
Spanish-American War, *33*
Spencer, Dave, 70
spiderwort (*Tradescantia ohiensis*), 103, 191
Spiranthes orchids, 30
spiritual connection to land, 272–73
Spring, Jim, 264–65
spring break, 73, 77
Spring Flora of Wisconsin (Fassett), 26, 271
spruce period, 183
Starker, Charles, 290–91
Starker, Clara, 290–91
Starker, Marie Runge, 290–91
Stevenson, Gordon, 267
Stevenson, Patricia, 227, 266–67
stewardship ethic, 276–77
stiff goldenrod (*Solidago rigida*), 133
Suevanna Prairie, 227, 232
sumacs (*Rhus typhina*), 133, 205, 226, 244
summer, 101–29
swamps, 207–8. *See also* marshes
swans, 71
sweet fern (*Comptonia peregrina*), 207
Swenson, Steve, 206, *233*, 239, 242–43
swimming, 7, *69*, 117–20, *121*
switchgrass (*Panicum virgatum*), 197

Taber, Richard, 251
tallgrass prairie, 229, 234, 247–48, *248*, 257

tamarack trees (*Larix laricina*), 96, 207–8
Telford, Jim, 137, 280
Temple, Stan, 238
Tenney, Jeanette, 28, 44
Tenney, Kent, 28, 44
Terbilcox, Frank, 201, 219, *233*
Texas prairies, 199
timber industry, 25–26
tiny sedges (*Carex* sp.), 206
tomatoes, 130
Tom's Meadow, 7
tool care, 75–76, *76*
topography of Shack area, *179*
tracking, 68
traguitos, 40, 77
trails, 127–29, 260–61
"Travels in Georgia" (McPhee), 100
tree house, 110–11
tree planting, 24–32, *28*, *29*, 73–77, 79–80, 95, 259–63. *See also specific tree species*
Trillium grandiflorum, 193
Tropical Forestry Initiative (TFI), 258–63
turkeyfoot bluestem (*Andropogon gerardii*), 197, 205
Turner, Fredrick Jackson, 28
Turner Ridge, 232
Two Bears Prairie, 174, 225, 232
two-man crosscut saw, 51, *54*, 55–56, 70–71, 75, 77, 115

Udall, Stewart, 251
University of California, Berkeley, 249–52
University of Wisconsin, xi, 125, 166, 227–28, 258, 293
University of Wisconsin Arboretum, 2, 189, 201, 215, 217, 224

upland game hunting, 173
US Fish and Wildlife Service, 254–55
US Forest Service, 252
US Geological Survey, 263, 273

vandalism, 60–61
Van Hoosen, Dorothy, 217
Van Hoosen, Russell, 210–11, 217
vegetable gardens, 9, 130–32, *131, 132*, 203–5
vegetation phases, 183–84
Vilas Park, 163, 164, *165, 167, 168*
vine bittersweet (*Celastrus scandens*), 193, 197

wahoo (*Euonomous afropurpureus*), 192–93
walleye pike, *122*
walnut (*Juglans nigra*), 228
warblers, 89–91, 124
Ward, Peter, 286
waterfowl, 141, 287
water lilies, *185*
water pump, 19, *20*, 38, *107*
The Way West (Guthrie), 83–85
Webster, Mr., 10, *11*
weeds
 and the Bradley Study Center, 219–20
 and Charlie Bradley's Woods and Prairie, 229
 and first visit to Shack property, 3, 5
 impact of agricultural practices, 188
 and prairie building, 224–25
 and reintroduction of natives, 195, 199
 and vegetable garden at Shack, 132–33
 and yard garden at Shack, 191, *192*
Wesson, Cynthia, 163, 164
West High School, 284–86
wetlands, 96, 230–32, *231*
Wet Prairie, 232
whiskey sours, 108
white birch (*Betula papyrifera*), 191–92. See also Birch Row
white oaks (*Quercus alba*), 186–88, *187*, 213–14, 218, *218*, 228–29
white pines (*Pinus strobus*), 24–32, 42, 73, 91, 183, 193–94, 208–9
whitewash, 44
white wild indigo (*Baptisia leucantha*), 197
Whitney, Joel, 295
whorled loosestrife (*Lysimachia quadrifolia*), 271
wild ginger, 201
wild onions (*Allium cernuum*), 205–6
willows, 115
Winkler, Barbara, 126, 183
Winnebago Indians, 5, 9
winter
 banding birds, 59–60
 and firewood, 51–60, *53–55, 57–58*, 69–72
 games, 67–69
 and the river, 62–67
 and vandalism of the Shack, 60–61
 and wildlife signs, 49–50, *52*
Wisconsin Archery Tournament, 163–65
Wisconsin Conservation District, 148
Wisconsin Department of Natural Resources, 238

Wisconsin Glaciation, *181*
Wisconsin River, 7
   beach, 7, *69*, 117–20, *121*
   and fishing, *64*
   and geology of the Shack area, 177
   and land restoration efforts, 230
   and location of Shack property, 1
   and neighbors of the Shack, 11–12
   and phenology observations, 237
   and restoration projects, 239
   river flotsam, 13–14, 66–67, 119
   and the Slough, 62–63
   and topography of Shack area, *179*
   in winter, 68, *69*
Wolf River Apples, 10, *11*, 130, 203
wood anemone (*Anemone quinquefolia*), 193

woodcocks, 87–88
woodcutting station, 51, *56*
wood ducks, 235, 236
wood flooring, 42–44
woodland thrushes, 90
woodpeckers, 25, 59, 91, 212, 252
woodworking/carpentry
   and archery, 157–60, *159*
   and Leopold benches, *20*, *28*, *105*, 111–17, *114*
   and reconstruction of the Shack, 12–24, *14*, *15*, *18*, *20*, *21*
World War II, 127
Wright, Vim, 266

yellow coneflowers, *221*
yellowthroat warbler, 89